Disclaimer

The publisher of this book is by no way associated with the National Institute of Standards and Technology (NIST). The NIST did not publish this book. It was published by 50 page publications under the public domain license.

50 Page Publications.

Book Title: Cell Phone Forensic Tools: An Overview and Analysis Update

Book Author: Richard P. Ayers; Wayne Jansen; Aurelien M. Delaitre; Ludovic Moenner

Book Abstract: Cell phones and other handheld devices incorporating cell phone capabilities (e.g., Personal Digital Assistant (PDA) phones) are ubiquitous. Rather than just placing calls, certain phones allow users to perform additional tasks such as SMS (Short Message Service) messaging, Multi-Media Messaging Service (MMS) messaging, IM (Instant Messaging), electronic mail, Web browsing, and basic PIM (Personal Information Management) applications (e.g., phone and date book). PDA phones, often referred to as smart phones, provide users with the combined capabilities of both a cell phone and a PDA. In addition to network services and basic PIM applications, one can manage more extensive appointment and contact information, review electronic documents, give a presentation, and perform other tasks. All but the most basic phones provide individuals with some ability to load additional applications, store and process personal and sensitive information independently of a desktop or notebook computer, and optionally synchronize the results at some later time. As digital technology evolves, the capabilities of these devices continue to improve rapidly. When cell phones or other cellular devices are involved in a crime or other incident, forensic examiners require tools that allow the proper retrieval and speedy examination of information present on the device. This report provides an overview on current tools (that have undergone significant updates or were not examined in NISTIR 7250: Cell Phone Forensic Tools: An Overview and Analysis) designed for acquisition, examination, and reporting of data discovered on cellular handheld devices, and an understanding of their capabilities and limitations.

Citation: NIST Interagency/Internal Report (NISTIR) - 7387

Keyword: cell phones; computer forensics; handheld devices; mobile devices

NISTIR 7387

Cell Phone Forensic Tools:
An Overview and Analysis Update

Rick Ayers
Wayne Jansen
Ludovic Moenner
Aurelien Delaitre

NISTIR 7387

Cell Phone Forensic Tools:
An Overview and Analysis Update

Rick Ayers
Wayne Jansen
Aurelien Delaitre
Ludovic Moenner

COMPUTER SECURITY

Computer Security Division
Information Technology Laboratory
National Institute of Standards and Technology
Gaithersburg, MD 20899-8930

March 2007

U.S. Department of Commerce
Carlos M. Gutierrez, Secretary

Technology Administration
Robert Cresanti, Acting Under Secretary of Commerce for Technology

National Institute of Standards and Technology
William A. Jeffrey, Director

Reports on Computer Systems Technology

The Information Technology Laboratory (ITL) at the National Institute of Standards and Technology (NIST) promotes the U.S. economy and public welfare by providing technical leadership for the Nation's measurement and standards infrastructure. ITL develops tests, test methods, reference data, proof of concept implementations, and technical analysis to advance the development and productive use of information technology. ITL's responsibilities include the development of technical, physical, administrative, and management standards and guidelines for the cost-effective security and privacy of sensitive unclassified information in Federal computer systems. This Interagency Report discusses ITL's research, guidance, and outreach efforts in computer security, and its collaborative activities with industry, government, and academic organizations.

National Institute of Standards and Technology Interagency Report
164 pages (2007)

Certain commercial entities, equipment, or materials may be identified in this document in order to describe an experimental procedure or concept adequately. Such identification is not intended to imply recommendation or endorsement by the National Institute of Standards and Technology, nor is it intended to imply that the entities, materials, or equipment are necessarily the best available for the purpose.

Abstract

Cell phones and other handheld devices incorporating cell phone capabilities (e.g., Personal Digital Assistant [PDA] phones) are ubiquitous. Rather than just placing calls, most phones allow users to perform additional tasks, including Short Message Service (SMS) messaging, Multi-Media Messaging Service (MMS) messaging, Instant Messaging (IM), electronic mail, Web browsing, and basic Personal Information Management (PIM) applications (e.g., phone and date book). PDA phones, often referred to as smart phones, provide users with the combined capabilities of both a cell phone and a PDA. In addition to network services and basic PIM applications, one can manage more extensive appointment and contact information, review electronic documents, give a presentation, and perform other tasks.

All but the most basic phones provide individuals with some ability to load additional applications, store and process personal and sensitive information independently of a desktop or notebook computer, and optionally synchronize the results at some later time. As digital technology evolves, the existing capabilities of these devices continue to improve rapidly. When cell phones or other cellular devices are involved in a crime or other incident, forensic examiners require tools that allow the proper retrieval and speedy examination of information present on the device. This report provides an overview on current tools designed for acquisition, examination, and reporting of data discovered on cellular handheld devices, and an understanding of their capabilities and limitations. It is a follow-on to *NISTIR 7250 Cell Phone Forensic Tools: An Overview and Analysis*[*], which focuses on tools that have undergone significant updates since that publication or were not covered previously.

---

[*] NISTIR 7250 – Cell Phone Forensic Tools: An Overview and Analysis can be accessed via: **http://csrc.nist.gov/publications/nistir/nistir-7250.pdf**

Purpose and Scope

The purpose of this report is to inform law enforcement, incident response team members, and forensic examiners about the capabilities of present day forensic software tools that have undergone significant updates or were not mentioned in the previous report *Cell Phone Forensic Tools: An Overview and Analysis*. These tools have the ability to acquire information from cell phones operating over Code Division Multiple Access (CDMA), Time Division Multiple Access (TDMA), Global System for Mobile communications (GSM) cellular networks and running various operating systems, including Symbian, Research in Motion (RIM), Palm OS, Pocket PC, and Linux.

An overview of each tool describes the functional scope and services provided for acquiring and analyzing evidence contained on cell phones and PDA phones. Generic scenarios were devised to mirror situations that arise during a forensic examination of these devices and their associated media. The scenarios are structured to reveal how selected tools react under various situations. Although generic scenarios were used in analyzing forensic tools, the procedures are not intended to serve as a formal product test or as a comprehensive evaluation. Additionally, no claims are made on the comparative benefits of one tool versus another. Instead, the report offers a broad and probing perspective on the state of the art of present-day forensic software tools for cell phones and PDA phones. Alternatives to using a forensic software tool for digital evidence recovery, such as desoldering and removing memory from a device to read out its contents or using a built-in hardware test interface to access memory directly, are outside the scope of this report.

It is important to distinguish this effort from the Computer Forensics Tool Testing (CFTT) project, whose objective is to provide measurable assurance to practitioners, researchers, and other users that the tools used in computer forensics investigations provide accurate results. Accomplishing this goal requires the development of rigorous specifications and test methods for computer forensics tools and the subsequent testing of specific tools against those specifications, which goes far beyond the analysis described in this document. The CFTT is the joint effort of the National Institute of Justice, the National Institute of Standards and Technology (NIST), the Office of Law Enforcement Standards (OLES), the U.S. Department of Defense, Federal Bureau of Investigation (FBI), U.S. Secret Service, the U.S. Immigration and Customs Enforcement (ICE), and other related agencies.[*]

The publication is not to be used as a step-by-step guide for executing a proper forensic investigation involving cell phones and PDA phones, or construed as legal advice. Its purpose is to inform readers of the various technologies available and areas for consideration when employing them. Before applying the material in this report, readers are advised to consult with management and legal officials for compliance with laws and regulations (i.e., local, state, federal, and international) that pertain to their situation.

---

[*] For more information on this effort see **www.cftt.nist.gov**

Audience

The primary audience of the Cell Phone Forensic Tool document is law enforcement, incident response team members, and forensic examiners who are responsible for conducting forensic procedures related to cell phone devices and associated removable media.

Table of Contents

INTRODUCTION 1

BACKGROUND 3

    SUBSCRIBER IDENTITY MODULE 5
    REMOVABLE MEDIA 6

FORENSIC TOOLKITS 8

    DEVICE SEIZURE 11
    PILOT-LINK 11
    GSM .XRY 12
    OXYGEN PHONE MANAGER 12
    MOBILEDIT! 13
    BITPIM 13
    TULP2G 13
    SECUREVIEW 14
    PHONEBASE2 14
    CELLDEK 14
    SIMIS2 14
    FORENSICSIM 14
    FORENSIC CARD READER 15
    SIMCON 15
    USIMDETECTIVE 15

ANALYSIS OVERVIEW 16

    TARGET DEVICES 16
    SCENARIOS 23

SYNOPSIS OF DEVICE SEIZURE 28

    POCKET PC PHONES 28
    PALM OS PHONES 29
    BLACKBERRY DEVICES 30
    CELL PHONES 32
    ACQUISITION STAGE 32
    SEARCH FUNCTIONALITY 35
    GRAPHICS LIBRARY 36
    BOOKMARKING 36
    ADDITIONAL TOOLS 37
    REPORT GENERATION 39
    SCENARIO RESULTS -PDA S 39
    SCENARIO RESULTS – CELL PHONES 40
    SCENARIO RESULTS -SIM CARD ACQUISITION 41

SYNOPSIS OF PILOT-LINK 42

SYNOPSIS OF GSM .XRY 43

- SUPPORTED PHONES 43
- ACQUISITION STAGE 43
- SEARCH FUNCTIONALITY 46
- GRAPHICS LIBRARY 46
- REPORT GENERATION 47
- SCENARIO RESULTS – CELL PHONES 50
- SCENARIO RESULTS -SIM CARD ACQUISITION 51

SYNOPSIS OF OXYGEN PHONE MANAGER 52

SYNOPSIS OF MOBILEDIT! 53
- SIM CARD ACQUISITION 54

SYNOPSIS OF BITPIM 54

SYNOPSIS OF TULP2G 55
- SIM CARD ACQUISITION 57

SYNOPSIS OF SECUREVIEW 58
- SUPPORTED PHONES 58
- ACQUISITION STAGE 58
- SEARCH FUNCTIONALITY 61
- GRAPHICS LIBRARY 61
- REPORT GENERATION 61
- SCENARIO RESULTS – CELL PHONES 61
- SCENARIO RESULTS -SIM CARD ACQUISITION 63

SYNOPSIS OF PHONEBASE2 65
- SUPPORTED PHONES 66
- ACQUISITION STAGE 66
- SEARCH FUNCTIONALITY 70
- GRAPHICS LIBRARY 70
- REPORT GENERATION 70
- SCENARIO RESULTS – CELL PHONES 71
- SCENARIO RESULTS -SIM CARD ACQUISITION 72

SYNOPSIS OF CELLDEK 73
- SUPPORTED PHONES 73
- ACQUISITION STAGE 73
- SEARCH FUNCTIONALITY 76
- GRAPHICS LIBRARY 77
- REPORT GENERATION 78
- SCENARIO RESULTS – CELL PHONES 78
- SCENARIO RESULTS -SIM CARD ACQUISITION 79

SYNOPSIS OF SIMIS2 81

SYNOPSIS OF FORENSICSIM 83

SYNOPSIS OF FORENSIC CARD READER 84

SYNOPSIS OF SIMCON	**85**
SYNOPSIS OF USIMDETECTIVE	**86**
ACQUISITION STAGE	86
SEARCH FUNCTIONALITY	87
GRAPHICS LIBRARY	87
REPORT GENERATION	87
SCENARIO RESULTS	90
CONCLUSIONS	**92**
GLOSSARY OF ACRONYMS	**93**
APPENDIX A: DEVICE SEIZURE RESULTS – SMART DEVICES	**96**
BLACKBERRY 7750	96
BLACKBERRY 7780	97
KYOCERA 7135	99
MOTOROLA MPX220	100
SAMSUNG I700	101
PALMONE TREO 600	102
APPENDIX B: DEVICE SEIZURE RESULTS – CELL PHONES	**105**
AUDIOVOX 8910	105
ERICSSON T68I	106
LG 4015	107
MOTOROLA C333	108
MOTOROLA V66	109
MOTOROLA V300	110
NOKIA 3390	111
NOKIA 6610I	112
SANYO PM-8200	113
APPENDIX C: GSM .XRY RESULTS	**115**
ERICSSON T68I	115
MOTOROLA C333	116
MOTOROLA V66	117
MOTOROLA V300	118
NOKIA 6610I	119
NOKIA 6200	120
NOKIA 7610	121
APPENDIX D: SECUREVIEW RESULTS	**123**
AUDIOVOX 8910	123
ERICSSON T68I	123
LG4015	124
MOTOROLA C333	125
MOTOROLA V66	126
MOTOROLA V300	127

  Nokia 6200... 128
  Sanyo PM-8200 129

APPENDIX E: PHONEBASE2 RESULTS.. 131

  Ericsson T68i 131
  Motorola V66 132
  Motorola V300 133
  Nokia 6610i 134

APPENDIX F: CELLDEK RESULTS 136

  Ericsson T68i 136
  Motorola MPx220 137
  Motorola v66 138
  Motorola v300 139
  Nokia 6200 140
  Nokia 6610i 141

APPENDIX G: SIM SEIZURE – EXTERNAL SIM RESULTS 143

  SIM 5343 143
  SIM 8778 143
  SIM 1144 144

APPENDIX H: GSM .XRY – EXTERNAL SIM RESULTS 145

  SIM 5343 145
  SIM 8778 145
  SIM 1144 145

APPENDIX I: SECUREVIEW – EXTERNAL SIM RESULTS 147

  SIM 5343 147
  SIM 8778 147
  SIM 1144 147

APPENDIX J: PHONEBASE2 – EXTERNAL SIM RESULTS 149

  SIM 5343 149
  SIM 8778 149
  SIM 1144 150

APPENDIX K: CELLDEK – EXTERNAL SIM RESULTS 151

  SIM 5343 151
  SIM 8778 151
  SIM 1144 152

APPENDIX L: USIMDETECTIVE – EXTERNAL SIM RESULTS 153

  SIM 5343 153
  SIM 8778 153
  SIM 1144 153

Acknowledgements

The authors, Rick Ayers, Wayne Jansen, Aurelien Delaitre, and Ludovic Moenner wish to express their gratitude to colleagues who reviewed drafts of this document. In particular, their appreciation goes to Karen Scarfone, Murugiah Souppaya and Tim Grance from NIST, Rick Mislan from Purdue University, and Lee Reiber from Mobile Forensics for their research, technical support, and written contributions to this document. The authors would also like to express thanks to all others who assisted with our internal review process.

This report was sponsored in part by Dr. Bert Coursey of the Department of Homeland Security (DHS). The Department's support and guidance in this effort are greatly appreciated.

Introduction

Computer forensics involves the identification, preservation, extraction, documentation, and analysis of computer data. Computer forensic examiners follow clear, well-defined methodologies and procedures that can be adapted for specific situations. Such methodologies consist of the following steps:

- Prepare a forensic copy (i.e., an identical bit-for-bit physical copy) of the acquired digital media, while preserving the acquired media's integrity.
- Examine the forensic copy to recover information.
- Analyze the recovered information and develop a report documenting any pertinent information uncovered.

Forensic toolkits are intended to facilitate the work of examiners, allowing them to perform the above steps in a timely and structured manner, and improve the quality of the results. This paper discusses available forensic software tools for handheld cellular devices, highlighting the facilities offered and associated capabilities.

Forensic software tools strive to address a wide range of applicable devices and handle the most common investigative situations with modest skill level requirements. These tools typically perform logical acquisitions using common protocols for synchronization, debugging, and communications. More complicated situations, such as the recovery of deleted data, often require highly specialized hardware-based tools and expertise, which is not within the scope of this report.

Handheld device forensics is a fairly new and emerging subject area within the computer forensics field, which traditionally emphasized individual workstations and network servers. Discrepancies between handheld device forensics and classical computer forensics exist due to several factors, including the following, which constrain the way in which the tools operate:

- The orientation toward mobility (e.g., compact size and battery powered, requiring specialized interfaces, media, and hardware)
- The filesystem residing in volatile memory versus non-volatile memory on certain systems
- Hibernation behavior, suspending processes when powered off or idle, but remaining active
- The diverse variety of embedded operating systems used
- Short product cycles for introducing new handheld devices

Most cell phones offer comparable sets of basic capabilities. However, the various families of devices on the marketplace differ in such areas as the hardware technology, advanced feature set, and physical format. This paper looks at forensic software tools for a number of popular platforms, including Symbian, Research In Motion (RIM), Pocket PC, and Palm OS devices. Together these platforms comprise the majority of the so-called smart phone devices currently available and in use. More basic phones, produced by various manufacturers and operational on various types of cellular networks are also addressed in the paper.

The remaining sections provide an overview of cell phones, memory cards, and forensic toolkits; describe the scenarios used to analyze the various tools and toolkits; present the findings from applying the scenarios; and summarize the conclusions drawn. The reader is assumed to have some background in computer forensics and technology. The reader should also be apprised that the tools discussed in the report are rapidly evolving, with new versions and better capabilities available regularly. The tool manufacturer should always be contacted for up-to-date information.

Background

Cell phones are highly mobile communications devices that can do an array of functions ranging from that of a simple digital organizer to that of a low-end personal computer. Designed for mobility, they are compact in size, battery powered, and lightweight, often use proprietary interfaces or operating systems, and may have unique hardware characteristics for product differentiation. Overall, they can be classified as basic phones that are primarily simple voice and messaging communication devices; advanced phones that offer additional capabilities and services for multimedia; and smart phones or high-end phones that merge the capabilities of an advanced phone with those of a Personal Digital Assistant (PDA).

Figure 1 gives an overview of the hardware characteristics of basic, advanced, and high-end cell phones for display quality, processing and storage capacity, memory and I/O expansion, built-in communications, and video and image capture. The bottom of the diagram shows the range of cellular voice and data advances from kilobit analog networks, still in use today, to megabit 3rd generation digital networks in the planning and early deployment stages. The diagram attempts to illustrate that more capable phones can capture and retain not only more information, but also more varied information, through a wider variety of sources, including removable memory modules, other wireless interfaces, and built-in hardware. Note that hardware components can and do vary from those assignments made in the diagram and, over time, technology once considered high end or advanced eventually appears in what would then be considered a basic phone.

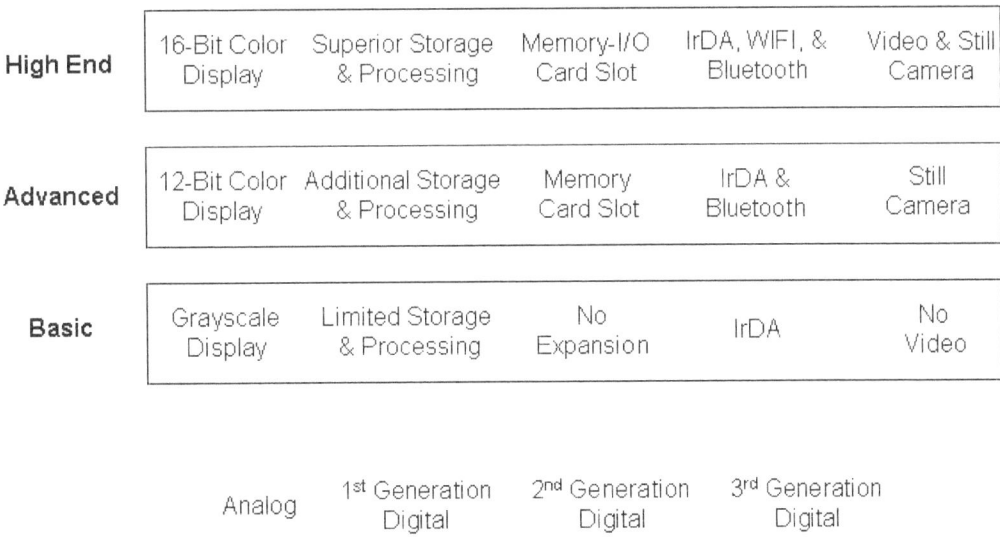

Figure 1: Phone Hardware Components

Just as with hardware components, software components involved in communications vary with the class of phone. Basic phones normally include text messaging using the Short Message Service (SMS). An advanced phone might add the ability to send simple picture messages or lengthy text messages using the Extended Message Service (EMS), while a high-end phone typically supports the Multimedia Message Service (MMS) to exchange sounds, color images, and text. Similarly, the ability to chat online directly with another user may be unsupported, supported through a dedicated SMS channel, or supported with a full Instant Messaging (IM) client. High-end phones typically support full function email and Web clients that respectively use Post Office Protocol (POP)/Internet Message Access Protocol (IMAP)/Simple Mail Transfer Protocol (SMTP) and HTTP, while advanced phones provide those services via Wireless Application Protocol (WAP), and basic phones do not include any support. Figure 2 gives an overview of the capabilities usually associated with each class of phone.

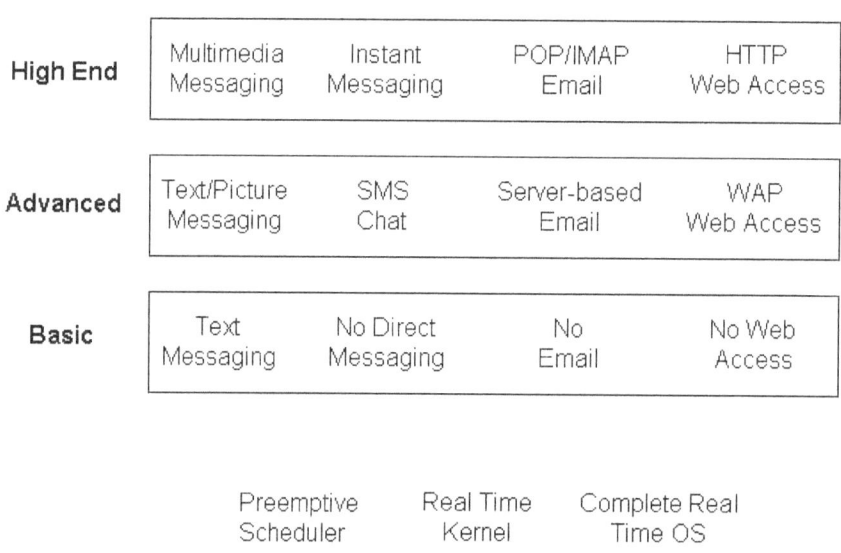

Figure 2: Phone Software Components

Most basic and many advanced phones rely on proprietary real-time operating systems developed by the manufacturer. Commercially embedded operating systems for cellular devices are also available that range from a basic preemptive scheduler with support for a few other key system calls to more sophisticated kernels with scheduling alternatives, memory management support, device drivers, and exception handling, to complete embedded real-time operating systems. The bottom of Figure 2 illustrates this range.

Many high-end smart phones have a PDA heritage, evolving from Palm OS and Pocket PC (also known as Windows mobile) handheld devices. As wireless telephony modules were incorporated into such devices, the operating system capabilities were enhanced to accommodate the functionality. Similarly, the Symbian OS used on many smart phones also stems from an electronic organizer heritage. RIM OS devices, which emphasize push technology for email messaging, are another device family that also falls into the smart phone category.

Subscriber Identity Module

Another useful way to classify cellular devices is by whether they involve a Subscriber Identity Module (SIM). A SIM is removable card designed for insertion into a device, such as a handset. SIMs originated with a set of specifications originally developed by the Conference of European Posts and Telecommunications (CEPT) and continued by the European Telecommunications Standards Institute (ETSI) for Global System for Mobile communications (GSM) networks. GSM standards mandate the use of a SIM for the operation of the phone. Without it, a GSM phone cannot operate. In contrast, present-day Code Division Multiple Access (CDMA) phones do not require a SIM. Instead, SIM functionality is incorporated directly within the device.

A SIM is an essential component of a GSM cell phone that contains information particular to the user. A SIM is a special type of smart card that typically contains between 16 to 64 kilobytes (KB) of memory, a processor, and an operating system. A SIM uniquely identifies the subscriber, determines the phone's number, and contains the algorithms needed to authenticate a subscriber to a network. A user can remove the SIM from one phone, insert it into another compatible phone, and resume use without the need to involve the network operator. The hierarchically organized filesystem of a SIM is used to store names and phone numbers, received and sent text messages, and network configuration information. Depending on the phone, some of this information may also coexist in the memory of the phone or reside entirely in the memory of the phone instead of the SIM. While SIMs are most widely used in GSM systems, compatible modules are also used in Integrated Digital Enhanced Network (IDEN) phones and Universal Mobile Telecommunications System (UMTS) user equipment (i.e., a Universal Subscriber Identity Module [USIM]). Because of the flexibility a SIM offers GSM phone users to port their identity and information between devices, eventually all cellular phones are expected to include SIM capability.

Though two sizes of SIMs have been standardized, only the smaller size shown at left is broadly used in GSM phones today. The module has a width of 25 mm, a height of 15 mm, and a thickness of .76 mm, which is roughly the size of a postage stamp. Its 8-pin connectors are not aligned along a bottom edge as might be expected, but instead form a circular contact pad integral to the smart card chip, which is embedded in a plastic frame. Also, the slot for the SIM card is normally not accessible from the exterior of the phone as with a memory card. When a SIM is inserted into a phone and pin contact is made, a serial interface is used to communicate with the computing platform using a half-duplex protocol. SIMs can be removed from a phone and read using a specialized SIM card reader and software. A SIM can also be placed in a standard-size smart card adapter and read using a conventional smart card reader.

As with any smart card, its contents are protected and a PIN can be set to restrict access. Two PINs exist, sometimes called PIN1 and PIN2 or CHV1 and CHV2. These PINs can be modified or disabled by the user. The SIM allows only a preset number of attempts, usually three, to enter the correct PIN before further attempts are blocked. Entering the correct PIN Unblocking Key (PUK) resets the PIN number and the attempt counter. The PUK can be obtained from the service provider or the network operator based on the SIM's identity (i.e., its Integrated Circuit Card Identifier [ICCID]). If the number of attempts to enter the PUK correctly exceeds a set limit, normally ten attempts, the card becomes blocked permanently.

Removable Media

Removable media extends the storage capacity of a cell phone, allowing individuals to store additional information beyond the device's built-in capacity. They also provide another avenue for sharing information between users that have compatible hardware. Removable media is non-volatile storage, able to retain recorded data when removed from a device. The main type of removable media for cell phones is a memory card. Though similar to SIMs in size, they follow a different set of specifications and have vastly different characteristics. Some card specifications also allow for I/O capabilities to support wireless communications (e.g., Bluetooth or WiFi) or other hardware (e.g., a camera) to be packaged in the same format.

A wide array of memory cards exists on the market today for cell phones and other mobile devices. The storage capacities of memory cards range from megabytes (MB) to gigabytes (GB) and come in sizes literally as small as a thumbnail. As technological advances continue, such media is expected to become smaller and offer greater storage densities. Fortunately, such media is normally formatted with a conventional filesystem (e.g., File Allocation Table [FAT]) and can be treated similarly to a disk drive, imaged and analyzed using a conventional forensic tool with a compatible media adapter that supports an Integrated Development Environment (IDE) interface. Such adapters can be used with a write blocker to ensure that the contents remain unaltered. Below is a brief overview of several commonly available types of memory cards used with cell phones.

Multi-Media Cards (MMC):[1]
A Multi-Media Card (MMC) is a solid-state disk card with a 7-pin connector. MMC cards have a 1-bit data bus. They are designed with flash technology, a non-volatile storage technology that retains information once power is removed from the card. Multi-Media Cards are about the size of a postage stamp (length-32 mm, width-24 mm, and thickness-1.4 mm). Reduced Size Multi-Media cards (RS-MMC) also exist. They are approximately one-half the size of the standard MMC card (length-18mm, width-24mm, and thickness-1.4mm). An RS-MMC can be used in a full-size MMC slot with a mechanical adapter. A regular MMC card can be also used in an RS-MMC card slot, though part of it will stick out from the slot. MMCplus and MMCmobile are higher performance variants of MMC and RS-MMC cards respectively that have 13-pin connectors and 8-bit data buses.

Secure Digital (SD) Cards:[2]
Secure Digital (SD) memory cards (length-32 mm, width-24 mm, and thickness-2.1mm) are comparable to the size and solid-state design of MMC cards. In fact, SD card slots can often accommodate MMC cards as well. However, SD cards have a 9-pin connector and a 4-bit data bus, which afford a higher transfer rate. SD memory cards feature an erasure-prevention switch; keeping the switch in the locked position protects data from accidental deletion. They also offer security controls for content protection (i.e., Content Protection Rights Management). MiniSD cards are an electrically compatible extension of the existing SD card standard in a more compact format (length-21.5 mm, width-20 mm, and

---

[1] Image courtesy of Lexar Media. Used by permission.
[2] Image courtesy of Lexar Media. Used by permission.

thickness-1.4 mm). They run on the same hardware bus as an SD card and also include content protection security features, but have a 11-pin connector and a smaller capacity potential due to size limitations. For backward compatibility, an adapter allows a MiniSD Card to work with existing SD card slots.

Memory Sticks:[3]

Memory sticks provide solid-state memory in a size similar to, but smaller than, a stick of gum (length-50mm, width-21.45mm, thickness-2.8mm). They have a 10-pin connector and a 1-bit data bus. As with SD cards, memory sticks also have a built-in erasure-prevention switch to protect the contents of the card. Memory Stick PRO cards offer higher capacity and transfer rates than standard Memory Sticks, using a 10-pin connector, but with a 4-bit data bus. Memory Stick Duo and Memory Stick PRO Duo, smaller versions of the Memory Stick and Memory Stick PRO, are about two-thirds the size of the standard memory stick (length-31mm, width-20mm, thickness-1.6mm). An adapter is required for a Memory Stick Duo or a Memory Stick PRO Duo to work with standard Memory Stick slots.

TransFlash:[4]

TransFlash, recently renamed MicroSD, is an extremely small size card (length-15 mm, width-11 mm, and thickness-1 mm). Because frequent removal and handling can be awkward, they are used more as a semi-removable memory module. TransFlash cards have an 8-pin connector and a 4-bit data bus. An adapter allows a TransFlash card to be used in SD-enabled devices. Similarly, the MMCmicro device is another ultra small card (length-14 mm, width-12 mm, and thickness-1.1 mm), compatible with MMC-enabled devices via an adapter. MMCmicro cards have a 11-pin connector and a 4-bit data bus. More recently, the Memory Stick Micro card has emerged, which is also ultra small (length-12.5 mm, width-15 mm, and thickness-1.2 mm) and, with an appropriate mechanical adaptor, able to be used in devices supporting fuller size cards in the Memory Stick family. Memory Stick Micro cards have an 11-pin connector and a 4-bit data bus

---

[3] Image courtesy of Lexar Media. Used by permission.
[4] Image courtesy of SanDisk. Used by permission.

Forensic Toolkits

The variety of forensic toolkits for cell phones and other handheld devices is diverse. A considerable number of software tools and toolkits exist, but the range of devices over which they operate is typically narrowed to distinct platforms for a manufacturer's product line, a family of operating systems, or a type of hardware architecture. Moreover, the tools require that the examiner have full access to the device (i.e., the device is not protected by some authentication mechanism or the examiner can satisfy any authentication mechanism encountered).

While most toolkits support a full range of acquisition, examination, and reporting functions, some tools focus on a subset. Similarly, different tools may be capable of using different interfaces (e.g., Infrared [IR], Bluetooth, or serial cable) to acquire device contents. The types of information a tool can acquire can range widely and include Personal Information Management (PIM) data (e.g., phone book); logs of phone calls; SMS/EMS/MMS messages, email, and IM content; URLs and content of visited Web sites; audio, video, and image content; SIM content; and uninterrupted image data. Information present on a cell phone can vary depending on several factors, including the following:
- The inherent capabilities of the phone implemented by the manufacturer
- The modifications made to the phone by the service provider or network operator
- The network services subscribed to and used by the user
- The modifications made to the phone by the user

Acquisition through a cable interface generally yields acquisition results superior to those from other device interfaces. However, although a wireless interface such as infrared or Bluetooth can serve as an alternative when the correct cable is not readily available, it should be used as a last resort due to the possibility of device modification during acquisition. Regardless of the interface used, one must be vigilant about any associated forensic issues. Note too that the ability to acquire the contents of a resident SIM may not be supported by some tools, particularly those strongly oriented toward PDAs. Table 1 lists open source and commercially available tools and the facilities they provide for certain types of cell phones.

Table 1: Cell Phone Tools

	Function	Features
Device Seizure	Acquisition, Examination, Reporting	• Targets Palm OS, Pocket PC, RIM OS phones and certain models of GSM, TDMA, and CDMA devices • Supports recovery of internal and external SIM • Supports only cable interface
pilot-link	Acquisition	• Targets Palm OS phones • Open source non-forensic software • No support for recovering SIM information • Supports only cable interface

	Function	Features
GSM .XRY	Acquisition, Examination, Reporting	• Targets certain models of GSM and CDMA phones • Internal and external SIM support • Requires PC/SC-compatible smart card reader for external SIM cards • Cable, Bluetooth, and IR interfaces supported • Supports radio-isolation SIM creation with proprietary card
Oxygen PM (forensic version)	Acquisition, Examination, Reporting	• Targets certain models of GSM phones • Supports only internal SIM acquisition
MOBILedit! Forensic	Acquisition, Examination, Reporting	• Targets certain models of GSM phones • Internal and external SIM support • Supports cable and IR interfaces
BitPIM	Acquisition, Examination	• Targets certain models of CDMA phones • Open source software with write-blocking capabilities • No support for recovering SIM information
TULP2G	Acquisition, Reporting	• Targets GSM and CDMA phones that use the supported protocols to establish connectivity • Internal and external SIM support • Requires PC/SC-compatible smart card reader for external SIM cards • Cable, Bluetooth, and IR interfaces supported • Supports radio-isolation SIM creation with GEM Xpresso card
SecureView	Acquisition, Examination, Reporting	• Targets GSM, CDMA and TDMA phones that use the supported protocols to establish connectivity • Internal and external SIM support • Requires PC/SC-compatible smart card reader for external SIM cards • Cable, Bluetooth, and IR interfaces supported
PhoneBase2	Acquisition, Examination, Reporting	• Targets GSM and CDMA phones that use the supported protocols to establish connectivity • External SIM support • Requires PC/SC-compatible smart card reader for external SIM cards • Cable, Bluetooth, and IR interfaces supported
CellDEK	Acquisition, Examination, Reporting	• Targets GSM and CDMA phones that use the supported protocols to establish connectivity • Internal and external SIM support • Built in PC/SC-compatible smart card reader for external SIM cards • Cable, Bluetooth and IR interfaces supported

Because of the way GSM phones are logically and physically partitioned into a handset and SIM, a number of forensic software tools have emerged that deal exclusively with SIMs independently of their handsets. The SIM must be removed from the phone and inserted into an appropriate reader for acquisition. SIM forensic tools require either a specialized reader that accepts a SIM directly or a general-purpose reader for a full-size smart card. For the latter, a standard-size smart card adapter is needed to house the SIM for use with the reader. Table 2 lists several SIM forensic tools. The first seven listed, Device Seizure, TULP2G, GSM .XRY, Mobiledit!, SecureView, PhoneBase2, and CellDEK also handle phone memory acquisition, as noted above.

Table 2: SIM Tools

	Function	Features
Device Seizure	Acquisition, Examination, Reporting	• Also recovers information from a SIM card via the handset • Requires Paraben's proprietary SIM reader
TULP2G	Acquisition, Reporting	• Also recovers information from a SIM card via the handset • Supports PC/SC reader • Supports radio-isolation SIM creation
GSM .XRY	Acquisition, Examination, Reporting	• Also recovers information from a SIM card via the handset • Supports PC/SC reader • Supports radio-isolation SIM creation
Mobiledit! Forensic	Acquisition, Examination, Reporting	• Also recovers information from a SIM card via the handset • Supports PC/SC reader
SecureView	Acquisition, Examination, Reporting	• Also recovers information from a SIM card via the handset • Supports PC/SC reader
PhoneBase2	Acquisition, Examination, Reporting	• External SIM cards only • Supports PC/SC reader
CellDEK	Acquisition, Examination, Reporting	• Also recovers information from a SIM card via the handset • Internal PC/SC reader
SIMIS2	Acquisition, Examination, Reporting	• External SIM cards only • Supports PC/SC reader • Supports radio-isolation SIM creation
ForensicSIM	Acquisition, Examination, Reporting	• External SIM cards only • Requires ForensicSIM's proprietary SIM reader • Supports radio-isolation SIM creation
Forensic Card Reader	Acquisition, Reporting	• External SIM cards only • Supports PC/SC reader
SIMCon	Acquisition, Examination, Reporting	• External SIM cards only • Supports PC/SC reader
USIMdetective	Acquisition, Examination, Reporting	• External SIM cards only • Supports PC/SC reader

Forensic software tools acquire data from a device in one of two ways: physical acquisition or logical acquisition. Physical acquisition implies a bit-by-bit copy of an entire physical store (e.g., a disk drive or RAM chip), while logical acquisition implies a bit-by-bit copy of logical storage objects (e.g., directories and files) that reside on a logical store. The difference lies in the distinction between memory as seen by a process through the operating system facilities (i.e., a logical view), versus memory as seen by the processor and other hardware components (i.e., a physical view). In general, physical acquisition is preferable, since it allows any data remnants present (e.g., unallocated RAM or unused filesystem space) to be examined, which otherwise would go unaccounted in a logical acquisition. Physical device images are generally more easily imported into another tool for examination and reporting. However, a logical acquisition provides a more natural and understandable organization of the information acquired. Thus, if possible, doing both types of acquisition is preferable.

Tools not designed specifically for forensic purposes are questionable and should be thoroughly evaluated before use. Although both forensic and non-forensic software tools generally use the same protocols to communicate with a device, non-forensic tools allow a two-way flow of information in order to populate and manage the device, and they usually do not compute hashes of acquired content for integrity purposes. Documentation also may be limited and source code unavailable for examination, respectively increasing the likelihood of error and decreasing confidence in the results. On the one hand, non-forensic tools might be the only means to retrieve information that could be relevant as evidence. On the other, they might overwrite, append, or otherwise cause information to be lost, if not used carefully.

The remainder of this chapter provides a brief introduction to each tool used for this report.

Device Seizure

Paraben's Device Seizure version 1.1 is a forensic software toolkit that allows forensic examiners to acquire, search, examine, and report data associated with PDAs running Palm OS, Windows CE, or RIM OS, and cell phones operating over CDMA, TDMA, and GSM networks and SIM cards via Paraben's proprietary RS-232 SIM reader. Device Seizure's features include the ability to perform a logical and physical acquisition (dependent upon the device type), providing a view of internal memory and relevant information concerning individual files and databases. To acquire data from cell phones using Paraben's Device Seizure software, the proper cable must be selected from either Paraben's Toolbox or a compatible cable (e.g., datapilot) to establish a data-link between the phone and the forensic workstation. The type of phone being acquired determines the cable interface. Serial RS-232 and USB data-link connections are established via the phone data port or the under-battery interface connection. Device Seizure uses the MD5 hash function to protect the integrity of acquired files. Additional features include bookmarking of information to be filtered and organized in a report format, searching for text strings within the acquired data, and automatically assembling found images under a single facility.

Pilot-Link

pilot-link is an open source software suite originally developed for the Linux community to allow information to be transferred between Linux hosts and Palm OS devices. It runs on several other desktop operating systems besides Linux, including Windows and Mac OS. About thirty command line programs comprise the software suite. To perform a physical and logical dump, pilot-link establishes a connection to the device with the aid of the Hotsync protocol. The two programs of interest to forensic examiners are pi-getram and pi-getrom, which respectively retrieve the physical contents of RAM and ROM from a device. Another useful program is pilot-xfer, which allows the installation of programs and the backup and restoration of databases. pilot-xfer provides a means to acquire the contents of a device logically. The contents retrieved with these utilities can be manually examined with the Palm OS Emulator (POSE), a compatible forensics tool such as EnCase, or a hex editor. pilot-link does not provide hash values of the information acquired, requiring a separate step to be carried out to obtain them.

GSM .XRY

Micro Systemation's SoftGSM .XRY is a forensic software toolkit for acquiring data from GSM, CDMA, 3G phones and SIM/USIM cards. The .XRY unit is able to connect to cell phone devices via IR, Bluetooth or a cable interface. After establishing connectivity, the phone model is identified with a corresponding picture of the phone, the device name, manufacturer, model, serial number (IMEI), Subscriber ID (IMSI), manufacturer code, device clock, and the PC clock. Data acquired from cell phone devices are stored in the .XRY format and cannot be altered, but can be exported into external formats and viewed with third-party applications. After a successful acquisition, the following fields may be populated with data, depending on the phone's functionality: Summary screen, Case data, General Information, Contacts, Calls, Calendar, SMS, Pictures, Audio, Files, Notes, Tasks, MMS, Network Information, and Video. Graphic files, audio files, and internal files present on the phone can be viewed internally or exported to the forensic workstation for safekeeping or further investigation.

Additionally, support exists for the creation of a substitute SIM for an original, using a proprietary rewritable SIM. The substitute SIM created provides the ability to acquire a device while disabling network communications. By providing radio isolation during acquisition, the SIM ID Cloner function eliminates the possibility of incoming data overwriting recoverable information. This added functionality also allows GSM devices that are not readable without a SIM present to be acquired if the SIM is missing.

Oxygen Phone Manager

The forensic version of Oxygen Phone Manager (OPM) is available for police departments, law enforcement units, and all government services that wish to use the software for investigative purposes. The forensic version differs from the non-forensic version of OPM by prohibiting any changes in data during acquisition. The previous report (*NISTIR 7250 Cell Phone Forensic Tools: An Overview and Analysis*) erroneously issued findings using the non-forensic version of Oxygen Phone Manager; therefore, each device was repopulated and re-examined using the forensic version of OPM. The scenario ratings earlier reported in *NISTIR 7250 Cell Phone Forensic Tools: An Overview and Analysis* are consistent with the findings in the present report using the forensic version of OPM.

Additionally, Oxygen Phone Manager provides a version of the software for Symbian based devices, allowing examiners to create a cable connection with the aid of an agent installed on the device, otherwise acquisitions must be peformed via Bluetooth or by manually browsing the device. Examiners are advised to take extreme precaution and consult with a forensic specialist when acquiring devices that require modification or Bluetooth connectivity, which may potentially corrupt valuable evidence present on the device.

OPM allows examiners to acquire data from the device and export the acquired data into multiple supported formats. The OPM software is tailored toward mobile phones and smart phones manufactured by: Nokia, Sony Ericsson, Siemens, Panasonic, Sendo, BenQ and some Samsung models. OPM provides software libraries, ActiveX libraries and components for Borland Delphi to software developers.

MOBILedit

MOBILedit! Forensic is an application that gives examiners the ability to acquire logically, search, examine and report data from GSM/CDMA/PCS cell phone devices. MOBILedit! is able to connect to cell phone devices via an Infrared (IR) port, a Bluetooth link, or a cable interface. After connectivity has been established, the phone model is identified by its manufacturer, model number, and serial number (IMEI) and with a corresponding picture of the phone. Data acquired from cell phone devices are stored in the .med file format. After a successful acquisition, the following fields are populated with data: subscriber information, device specifics, Phonebook, SIM Phonebook, Missed Calls, Last Numbers Dialed, Received Calls, Inbox, Sent Items, Drafts, Files folder. Items present in the Files folder, ranging from Graphics files to Camera Photos and Tones, depend on the phone's capabilities. Additional features include the myPhoneSafe.com service, which provides access to the IMEI database to register and check for stolen phones.

BitPIM

BitPIM is a phone management program that runs on Windows, Linux and Mac OS and allows the viewing and manipulation of data on cell phones. This data includes the phone book, calendar, wallpapers, ring tones, videos, memo, SMS, callhistory, T9 Database and the embedded filesystem. To acquire data successfully using BitPIM, examiners must have the proper driver and cable to form a connection between the phone and the forensic workstation. BitPIM provides detailed information contained in the help file, outlining supported phones, suggested cables to use with specific phone models, and notes and How-Tos about specific situations. BitPIM is distributed as open source software under the GNU General Public License.

TULP2G

TULP2G (2nd generation) is an open source forensic software tool originated by the Netherlands Forensic Institute that allows examiners to extract and read data from mobile cell phones and SIMs. TULP2G requires a forensic workstation running either Windows 2000 or XP, preferably with the latest patches and service pack installed, along with .NET 1.1 SP1. In order to take advantage of newly released 1.1 plug-ins, Windows XP SP2 is required. TULP2G acquires data from mobile phones using a proper data cable, Bluetooth or IrDA connection and a compatible protocol plug-in. Reading SIMs requires a PC/SC-compatible smart card reader and possibly an adapter to convert a small-sized SIM to the standard-size smart card format. Support for the creation of a substitute SIM, using the SIMIC plug-in and a GEM Xpresso test SIMs, provides the ability to acquire devices while disabling network communications,

SecureView

Susteen's SecureView is a forensic software toolkit that acquires data from mobile devices operating over GSM, CDMA, TDMA networks and extract data from SIM cards. SecureView provides examiners with a read-only secure environment, eliminating accidental manipulation or deletion of critical data, while supporting over 350 U.S. and Canadian phone models. SecureView acquires data from mobile phones via data cable, Bluetooth or an IrDA connection. Susteen's unique universal cellular cable kit provides a simple solution for multiple phone models and can be used with various mobile acquisition applications.

PhoneBase2

Envisage Systems Ltd. PhoneBase2 provides examiners the ability to securely acquire, search, examine and create finalized reports for data residing on devices operating over GSM and non-GSM networks as well as extract data from SIM cards. PhoneBase2 does not include data cables required to create a successful data-link connection from the device to PC. Various third-party hardware solutions (e.g., Susteen) can be utilized while the appropriate cable and driver are installed for successful communication via a cable interface. PhoneBase2 acquires data from mobile phones via data cable, Bluetooth or an IrDA interface

CellDEK

CellDEK from Logicube Inc. is designed to acquire data from cell phones operating over GSM and non-GSM networks, PDAs, SIM cards and flash-based media. The CellDEK unit provides examiners with the ability to connect to the aforementioned devices via a cable, Bluetooth or IrDA connection. The CellDEK terminal contains an embedded touch-screen PC, data cables for various phone manufacturers, PC/SC SIM card reader and a write-protected flash memory card reader, packaged in a rugged, watertight carrying case. Acquisitions are stored on the CellDEK's hard disk and can be moved or backed up to a USB thumb drive.

SIMIS2

SIMIS2 is a forensic tool from Crownhill USA that gives examiners the ability to extract data from a SIM securely and protect data integrity with cryptographic hashes. A USB dongle is needed to operate the software on a desktop computer. The SIMIS2 desktop is capable of decoding unicode data found on the SIM card, including active and deleted text messages and phone book information. The company also offers the SIMIS2 Mobile Handheld Reader, which is a portable standalone SIM reader that can capture SIM data for transfer to the SIMIS2 desktop. Support for the creation of a radio-isolation card, provides examiners with the ability to acquire devices without network interruption, via the SIMIS handheld unit.

ForensicSIM

Radio Tactic's ForensicSIM Toolkit consists of the following components: acquisition terminal, control card, data storage cards, analysis application, and card reader. The acquisition terminal is a standalone unit that guides the examiner through each step of the acquisition process. The ForensicSIM toolkit deals with two processes: acquisition of data and analysis of data. Data acquisition is carried out using the acquisition terminal. Data analysis is carried out using the ForensicSIM card reader, attached to a PC running the ForensicSIM analysis application. The terminal's primary function is to capture copies of the data from the target SIM to a set of data storage cards. A control card is used to provide the examiner access to the acquisition terminal, thwarting unauthorized use. The data storage cards consist of a master data storage card, a prosecution data storage card, a defense data storage card, and a handset access card. The handset access card serves as a substitute SIM for a phone, allowing the device to be acquired with network communications disabled. The toolkit allows examiners read-only access to SIMs and generates textual reports based on the contents acquired. Reports can be viewed internally, saved to disk, or printed for presentation purposes.

Forensic Card Reader

The Forensic Card Reader (FCR) consists of a USB smart card reader and the FCR software that gives examiners the ability to acquire data from SIM cards without modification. The examiner has the ability to select specific data elements that can be later stored and displayed in a finalized report. Operations details like case number, evidence number, and examiner can be automatically merged into the report and its filename. All usual data elements are acquired (e.g., phone directory, abbreviated dialing numbers, fixed dialing numbers and SMS messages), as well as the identifiers of the SIM and the subscriber. Special elements such as deleted SMS messages can also be acquired. The FCR stores a complete report in an XML format. SIM cards for GSM mobiles and 3G mobiles can be used with the FCR. Extended phone book entries can be acquired, including additional numbers and email addresses. The supplied FCR reader accepts either small or large SIM cards without the need for an adapter.

SIMCon

SIMCon works with any standard smart card reader compliant with the PC/SC standard. Upon completing the acquisition of the SIM card data, SIMCon card content is stored in unique files identified by a two-byte File ID code. Individual files may contain many informative elements called "items" and are displayed in tabular form. Each item, when selected, can be shown in hexadecimal or a textual interpretation. Besides standard SIM file content, SIMCon also has an option to do a comprehensive scan of all directories and files that may be present on the SIM, to acquire non-standardized directories and files. Examiners can create customized reports by selecting file information that pertains to the investigation.

USIMdetective

Quantaq Solutions' USIMdetective SIM acquisition tool provides examiners with the abiltity to acquire, examine and produce reports from any SIM or USIM card using a PC/SC compatible reader. Acquired elements can be displayed in a textual or hexadecimal format. Image Integrity Check (.iic) files are created with each acquisitions providing protected against data tampering during further examination and analysis. USIMdetective provides multiple report outputs types, from a "Standard Report" to a "File Content Report", which provides finer detail of acquired data.

Analysis Overview

A simple methodology was followed to understand and gauge the capabilities of the forensic tools described in the previous section. The main steps are illustrated in Figure 3. First, a set of target devices ranging from simple to smart phones was assembled. Then a set of prescribed activities, such as placing and receiving calls, was performed for each phone. After applying one or more of such scenarios, the contents of the phone and/or associated SIM were acquired using an available tool and examined to determine whether evidence of an activity could be recovered as expected. Finally, an assignment was made about how well the tool met predefined expectations. At least two different individuals performed each scenario and assigned a rating separately; any noted inconsistencies were resolved.

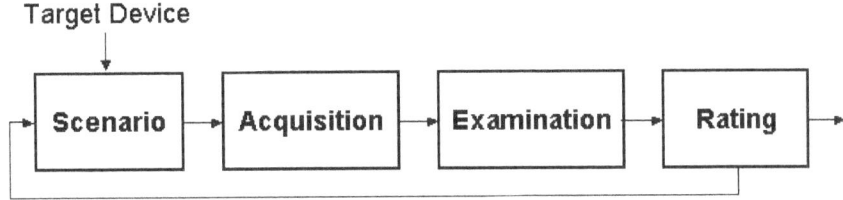

Figure 3: Tool Assessment

For GSM phones, two sets of scenarios were applied: one for handsets containing an associated SIM, and the other for SIMs removed from their handsets and examined independently. For CDMA and other types of phones that do not depend on a SIM, only the former set was used.

Target Devices

A suitable but limited number of target devices were needed on which to conduct the scenarios. The target devices selected, while not extensive, cover a range of operating systems, processor types, and hardware components. These variations were intended to uncover subtle differences in the behavior of the forensic tools in acquisition and examination. Table 3 highlights the key characteristics of each target device, listed roughly from devices with more capabilities to less-capable devices, rather than alphabetically. Note that the more capable devices listed have a PDA heritage, insofar as they use Windows Mobile, Palm OS, RIM OS, or Symbian operating systems.

Table 3: Target Device Characteristics

	Software	Hardware	Wireless
Samsung SGH-i300	Windows Mobile 2003 SE, SMS, EMS, MMS, Email (IMAP4, POP3) Web (HTML, WAP 2.0)	416 MHz Intel XScale processor 3 GB HDD 64 MB RAM Color display 262,144 color TFT display 1.3 megapixel / flash camera TransFlash slot	GSM 900/1800/1900 GPRS Bluetooth IrDA

	Software	Hardware	Wireless
Motorola MPX220	Windows Mobile for Smart Phones 2003 SMS, EMS, MMS SMS Chat Email (IMAP4, POP3) Web (HTML, WAP 2.0)	200 MHz OMAP 1611 processor 64 MB ROM 32 MB RAM Color display 2nd monochrome display camera MiniSD slot	GSM 850/900/ 1800/1900 GPRS Bluetooth IrDA
Treo 600	Palm OS 5.2 SMS, EMS, MMS SMS Chat Email (POP3, SMTP) Web (HTML 4.0, XHTML, WML 1.3)	144 MHz OMAP 1510 ARM-based processor 32 MB RAM (24 MB available) Color display QWERTY keypad SD/MMC slot (with SDIO)	GSM 850/900/ 1800/1900 GPRS IrDA
Sony Ericsson P910a	Symbian 7.0, UIQ 2.1 SMS, EMS, MMS Email (POP3, IMAP4) Web (WAP)	ARM 9 processor 64MB ROM 32MB RAM Color display Camera Memory Stick Duo Pro slot	GSM 850/1800/1900 HSCSD, GPRS Bluetooth IrDA
Samsung i700	Pocket PC 2002 Phone Edition SMS (no EMS/MMS) Email Web Instant messaging	300 MHz StrongArm PXA250 processor 32MB flash memory 64MB SDRAM Color display Swivel camera SD/MMC slot (with SDIO)	AMPS 800 CDMA 800/1900 1xRTT IrDA
Nokia 7610	Symbian 7.0, Series 60 2.0 SMS, MMSConcatenated SMS Email (SMTP, POP3, IMAP4) Instant messaging Web (WAP 2.0, HTML, XHTML and WML)	123 MHz processor 8 MB internal dynamic memory Color display Camera Reduced size MMC slot	GSM 850/1800/1900 HSCSD, GPRS Bluetooth
Kyocera 7135	Palm OS 4.1 SMS, EMS (no MMS) Email (POP, IMAP, SMTP) Web (HTML 3.2)	33 MHz Dragonball VZ processor 16 MB volatile Color display SD/MMC slot (with SDIO)	AMPS 800 CDMA 800/1900 1xRTT IrDA

	Software	Hardware	Wireless
BlackBerry 7780	RIM OS SMS Email (POP3) Web (WAP)	16 MB flash memory plus 2 MB SRAM Color display QWERTY keypad	GSM 850/1800/1900 GPRS
BlackBerry 7750	RIM OS SMS (no EMS/SMS) Email (POP3, IMAP4) Web (WAP 2.0, WML/HTML)	ARM7TDMI (Qualcomm 5100 chipset) 14 MB flash memory 2 MB SRAM Color display QWERTY keypad	CDMA 800/1900 1xRTT
Motorola V300	SMS, EMS, MMS SMS Chat Nokia Smart Message Instant messaging Email (SMTP, POP3, IMAP4) Web (WAP 2.0)	5 MB internal memory Color display Camera	GSM 900/1800/1900 GPRS
LG4015	SMS, EMS, MMS SMS Chat, Email Web (WAP 2.0)	610KB internal memory Color display	GSM 850/1900 GPRS
Nokia 6610i	Series 40 SMS, MMS Concatenated SMS SMS Chat No email Web (WAP 1.2.1 XHTML)	4 MB user memory 8-line color display Camera FM radio	GSM 900/1800/1900 HSCSD, GPRS IrDA
Ericsson T68i	SMS/EMS messaging MMS messaging Email (POP3,SMTP) SMS Chat Web (WAP 1.2.1/2.0, WLTS)	Color display Optional camera attachment	GSM 900/1800/1900 HSCSD, GPRS Bluetooth IrDA
Sanyo 8200	SMS, EMS Picture Mail Email Web WAP 2.0 Mobile-to-mobile (walkie talkie)	Color display 2nd color display Camera	AMPS 850 CDMA 850/1900

	Software	Hardware	Wireless
Nokia 6200	SMS, EMS, MMS Email over SMS SMS Chat Web (WAP 1.2.1, XHTML)	Color display FM radio	GSM 850/1800/1900 GPRS, EDGE IrDA
Audiovox 8910	EMS, MMS SMS Chat No email Web (WAP 2.0)	Color display 2nd monochrome display Camera	AMPS 850 CDMA800/1900 1xRTT
Motorola C333	SMS, EMS SMS chat Web (WAP 1.2.1)	Monochrome graphic display	GSM 850/1900 GPRS
Motorola V66	SMS (no EMS) AOL Instant Messenger Web (WAP 1.1)	Monochrome graphic display	GSM 900/1800/1900 GPRS
Nokia 3390	SMS Picture messaging Email over SMS AOL Instant Messager	Monochrome graphic display	GSM 1900

Not every tool supports every target device. In fact, the opposite is true – a specific tool typically supports only a limited number of devices. The determination of which tool to use for which device was based primarily on the tool's documented list of supported phones. Whenever ambiguity existed, an acquisition attempt was conducted to make a determination. Table 4 summarizes the various target devices used with each tool. The order of the devices bears no relevance on capabilities they are alphabetized for consistency throughout the rest of the document. The table excludes forensic SIM tools, which support most SIMs found in GSM devices.

Table 4: Target Devices Supported by Each Tool

	Device Seizure	Pilot-link	GSM .XRY	OPM	Mobiledit!	TULP2G	BitPIM	SecureView	PhoneBase2	CellDEK
Audiovox 8910	X						X	X[*]		
Blackberry 7750	X									
Blackberry 7780	X									
Ericsson T68i	X	X	X	X	X					

[*] Acquisition is supported only for non pay-as-you go carriers

	Device Seizure	Pilot-link	GSM .XRY	OPM	Mobiledit!	TULP2G	BitPIM	SecureView	PhoneBase2	CellDEK
Kyocera 7135	X	X								
LG4015	X							X*		
Motorola C333	X	X	X							
Motorola MPX220	X									
Motorola V66	X	X	X	X	X	X				
Motorola V300	X	X	X	X	X	X				
Nokia 3390	X			X				X		
Nokia 6200			X	X	X					
Nokia 6610i	X	X	X	X	X	X	X			
Nokia 7610			X	X						
Samsung i700	X									
Samsung SGH-i300										
Sanyo 8200	X							X		
Sony Ericsson P910a						X				
Treo 600	X	X								

Though SIMs are highly standardized, their content can vary among network operators and service providers. For example, a network operator might create an additional file on the SIM for use in its operations or might install an application to provide a unique service. SIMs may also be classified according to the "phase" of the GSM standards that they support. The three phases defined are phase 1, phase 2, and phase 2+, which correspond roughly to first, second, and 2.5 generation network facilities. Another class of SIMs in early deployment is Universal SIMs (USIMS) used in third generation (3G) networks.

Except for pay-as-you-go phones, each GSM phone was matched with a SIM that offered services compatible with the phone's capabilities. Only a subset of the SIMs used in the phone scenarios were used for the SIM scenarios. Table 5 lists the identifier and phase of the SIMs used in that analysis, the associated network operator, and some of the associated network

---

* Acquisition is supported only for non pay-as-you go carriers

services activated on the SIM. Except for pay-as-you-go phones, each GSM phone was matched with a SIM that offered services compatible with the phone's capabilities.

Table 5: SIMs

SIM	Phase	Network	Services
1144	2 - profile download required	AT&T	Abbreviated Dialing Numbers (ADN) Fixed Dialing Numbers (FDN) Short Message Storage (SMS) Last Numbers Dialed (LND) General Packet Radio Service (GPRS)
8778	2 - profile download required	Cingular	Abbreviated Dialing Numbers (ADN) Fixed Dialing Numbers (FDN) Short Message Storage (SMS) Last Numbers Dialed (LND) Group Identifier Level 1 (GID1) Group Identifier Level 2 (GID2) Service Dialing Numbers (SDN) General Packet Radio Service (GPRS)
5343	2 - profile download required	T-Mobile	Abbreviated Dialing Numbers (ADN) Fixed Dialing Numbers (FDN) Short Message Storage (SMS) Last Numbers Dialed (LND) General Packet Radio Service (GPRS)

Overall, SIM forensic tools do not recover every possible item on a SIM. While a few tools aim to recover all information present, most concentrate on a subset considered most useful as forensic evidence. The breadth of coverage varies considerably among tools. Table 6 entries provide an overview of those items recovered, listed at the left, by the various SIM forensic tools, listed across the top.

Table 6: Content Recovery Coverage

	Device Seizure	GSM .XRY	Mobiledit!	TULP2G	FCR	ForensicSIM	SIMCon	SIMIS2	SecureView	PhoneBase2	CellDEK	USIMdetective
International Mobile Subscriber Identity – IMSI	X	X	X	X	X	X	X			X	X	X
Integrated Circuit Card Identifier – ICCID	X	X	X	X	X	X	X			X	X	X
Mobile Subscriber ISDN – MSISDN	X	X		X	X	X	X			X	X	X
Service Provider Name – SPN	X			X	X	X	X	X	X			
Phase Identification – Phase	X	X	X			X	X	X	X	X		
SIM Service Table – SST				X	X	X	X					X
Language Preference – LP	X			X	X	X	X					X
Abbreviated Dialing Numbers – ADN	X	X	X	X	X	X	X	X	X	X	X	
Last Numbers Dialed – LND	X	X	X	X	X	X	X			X	X	X
Short Message Service – SMS												
• Read/Unread	X	X	X	X	X	X	X	X		X	X	X
• Deleted	X	X		X		X	X	X		X	X	X
PLMN selector – PLMNsel	X			X	X	X	X					X
Forbidden PLMNs – FPLMNs	X			X	X	X	X					X
Location Information – LOCI	X	X		X	X	X		X	X	X	X	
GPRS Location Information – GPRSLOCI	X					X	X	X				X

Scenario

The scenarios define a set of prescribed activities used to gauge the capabilities of the forensic tool to recover information from a phone, beginning with connectivity and acquisition and moving progressively toward more interesting situations involving common applications, file formats, and device settings. The scenarios are not intended to be exhaustive or to serve as a formal product evaluation. However, they attempt to cover a range of situations commonly encountered when examining a device (e.g., data obfuscation, data hiding, data purging) and are useful in determining the features and functionality afforded an examiner.

Table 7 gives an overview of these scenarios, which are generic to all devices that have cellular phone capabilities. For each scenario listed, a description of its purpose, method of execution, and expected results are summarized. Note that the expectations are comparable to those an examiner would have when dealing with the contents of a hard disk drive as opposed to a PDA/cell phone. Though the characteristics of the two are quite different, the recovery and analysis of information from a hard drive is a well-understood baseline for comparison and pedagogical purposes. Moreover, comparable means of digital evidence recovery from most phones exist, such as desoldering and removing non-volatile memory and reading out the contents with a suitable device programmer. Also note that none of the scenarios attempt to confirm whether the integrity of the data on a device is preserved when applying a tool – that topic is outside the scope of this document.

Table 7: Phone Scenarios

Scenario	Description
Connectivity and Retrieval	Determine whether the tool can successfully connect to the device and retrieve content from it. • Enable user authentication on the device before acquisition, requiring a PIN, password, or other known authentication information to be supplied for access. • Initiate the tool on a forensic workstation, attempt to connect with the device and acquire its contents, verify that the results are consistent with the known characteristics of the device. • Expect that the authentication mechanism(s) can be satisfied without affecting the tool, and information residing on the device can be retrieved.
PIM Applications	Determine whether the tool can find information, including deleted information, associated with Personal Information Management (PIM) applications such as phone book and date book. • Create various types of PIM files on the device, selectively delete some entries, acquire the contents of the device, locate and display the information. • Expect that all PIM-related information on the device can be found and reported, if not previously deleted. Expect that remnants of deleted information can be recovered and reported.

Dialed/Received Phone Calls	Determine whether the tool can find dialed and received phone calls, including unanswered and deleted calls. • Place and receive various calls to and from different numbers, selectively delete some entries, acquire the contents of the device, locate and display dialed and received calls. • Expect that all dialed and received phone calls on the device can be recognized and reported, if not previously deleted. Expect that remnants of deleted information can be recovered and reported.
SMS/MMS Messaging	Determine whether the tool can find placed and received SMS/MMS messages, including deleted messages. • Place and receive both SMS and MMS messages, selectively delete some messages, acquire the contents of the device, locate and display all messages. • Expect that all sent and received SMS/MMS messages on the device can be recognized and reported, if not previously deleted. Expect that remnants of deleted information can be recovered and reported.
Internet Messaging	Determine whether the tool can find sent and received email and Instant Message (IM) messages, including deleted messages. • Send and receive both IM and email messages, selectively delete some messages, acquire the contents of the device, locate and display all messages. • Expect that all sent and received IM and messages on the device can be recognized and reported, if not previously deleted. Expect that remnants of deleted information can be recovered and reported.
Web Applications	Determine whether the tool can find a visited Web site and information exchanged over the internet. • Use the device to visit specific Web sites and perform queries, selectively delete some data, acquire the contents of the device locate and display the URLs of visited sites and any associated data acquired (e.g., images, text). • Expect that information about most recent Web activity can be found and reported.
Text File Formats	Determine whether the tool can find and display a compilation of text files residing on the device, including deleted files. • Load the device with various types of text files (via email and device synchronization protocols) selectively delete some files, acquire the contents of the device, find and report the data. • Expect that all files with common text file formats (i.e., `.txt`, `.doc`, `.pdf`) can be found and reported, if not deleted. Expect that remnants of deleted information can be recovered and reported.

Graphics File Formats	Determine whether the tool can find and display a compilation of the graphics formatted files residing on the device, including deleted files. • Load the device with various types of graphics files, (via email and device synchronization protocols) selectively delete some files, acquire the contents of the device, locate and display the images. • Expect that all files with common graphics files formats (i.e., .bmp, .jpg, .gif, .tif, and .png) can be found, reported, and collectively displayed, if not deleted. Expect that remnants of deleted information can be recovered and reported.
Compressed Archive File Formats	Determine whether the tool can find text, images, and other information located within compressed-archive formatted files (i.e., .zip, .rar, .tar, .tgz, and self-extracting .exe) residing on the device. • Load the device with various types of file archives (via email and device synchronization protocols) acquire the contents of the device, find and display selected filenames and file contents. • Expect that text, images, and other information contained in the compressed archive formatted files can be found and reported.
Misnamed Files	Determine whether the tool can recognize file types by header information instead of file extension, and find common text and graphics formatted files that have been misnamed with misleading extensions. • Load the device (via email and device synchronization protocols) with various types of common text (e.g., .txt) and graphics files (e.g., .bmp, .jpg, .gif, and .png) that have been purposely misnamed, acquire the contents of the device, locate and display selected text and images. • Expect that all misnamed text and graphics files residing on the device can be recognized, reported, and, for images, displayed.
Peripheral Memory Cards	Determine whether the tool can acquire individual files stored on a memory card inserted into the device and whether deleted files can be identified and recovered. • Insert a formatted memory card containing text, graphics, archive, and misnamed files into an appropriate slot on the device, delete some files, acquire the contents of the device, find and display selected files and file contents, including deleted files. • Expect that the files on the memory card, including deleted files, can be properly acquired, found, and reported in the same way as expected with on-device memory.
Acquisition Consistency	Determine whether the tool provides consistent hashes on files resident on the device for two back-to-back acquisitions • Acquire the contents of the device and create a hash over the memory, for physical acquisitions, and over individual files, for logical acquisitions. • Expect that hashes over the individual file hashes are consistent between the two acquisitions, but inconsistent for the memory hashes.

Cleared Devices	Determine whether the tool can acquire any user information from the device or peripheral memory after a hard reset has been performed. • Perform a hard reset on the device, acquire its contents, and find and display previously available filenames and file contents. • Expect that no user files, except those contained on a peripheral memory card, if present, can be recovered.
Power Loss	Determine if the tool can acquire any user information from the device after it has been completely drained of power. • Completely drain the device of power by exhausting the battery or removing the battery overnight and then replacing, acquire device contents, and find and display previously available filenames and file contents. • Expect that no user files, except those contained on a peripheral memory card, if present, can be recovered.

A distinct set of scenarios was developed for SIM forensic tools. The SIM scenarios differ from the phone scenarios in several ways. SIMs are highly standardized devices with relatively uniform interfaces, behavior, and content. All of the SIM tools broadly support any SIM for acquisition via an external reader. Thus, the emphasis in these scenarios is on loading the memory of the SIM with specific kinds of information for recovery, rather than the memory of the handset. Once a scenario is completed using a suitable GSM phone or SIM management program, the SIM can be processed by each of the SIM tools in succession. Table 8 gives an overview of the SIM scenarios, including their purpose, method of execution, and expected results.

Table 8: SIM Scenarios

Scenario	Description
Basic Data	Determine whether the tool can recover subscriber (i.e., IMSI, ICCID, SPN, and LP elementary files), PIM (i.e., ADN elementary file), call (i.e., LND elementary file), and SMS message related information on the SIM, including deleted entries, and whether all of the data is properly decoded and displayed. • Populate the SIM with known PIM, call, and SMS message related information that can be verified after acquisition; then remove the SIM for acquisition and analysis. • Expect that all information residing on the SIM can be successfully acquired and reported.
Location Data	Determine whether the tool can recover location-related information (i.e., LOCI, LOCIGPRS, and FPLMN elementary files) on the SIM and whether all of the data is properly decoded and displayed. Location information can indicate where the device was last used for a particular service and other networks it might have encountered. • Register location-related data maintained by the network on the SIM by performing voice and data operations at known locations, then remove the SIM for acquisition and analysis. • Expect that all location-related information can be successfully acquired and reported.

Scenario	Description
EMS Data	Determine whether the tool can recover EMS messages over 160 characters in length and containing non-textual content, and whether all of the data is properly decoded and displayed for both active and deleted messages. EMS messages can convey pictures and sounds, as well as formatted text, as a series of concatenated SMS messages. • Populate the SIM with known EMS content that can be verified after acquisition; then remove the SIM for acquisition and analysis. • Expect that EMS messages can be successfully acquired and reported.
Foreign Language Data	Determine whether the tool can recover SMS messages and PIM data from the SIM that are in a foreign language, and whether all of the data is properly decoded and displayed. • Populate the SIM with known SMS and PIM content that can be verified after acquisition; then remove the SIM for acquisition and analysis. • Expect that foreign language data can be successfully acquired and reported.

The chapters that follow provide a brief discussion on tools previously covered in *NISTIR 7250 – Cell Phone Forensic Tools: An Overview and Analysis* and a detailed synopsis on tools not previously covered as well as ones that have undergone significant updates. A summary of the results of applying the above scenarios to the target devices determines the extent to which a given tool meets the expectations listed. The tool synopsis concentrates on several core functional areas: acquisition, search, graphics library, and reporting, and also other useful features such as tagging uncovered evidence with a bookmark.

The scenario results for each tool are weighed against the predefined expectations defined above in Table 7 and Table 8, and assigned a ranking. The entry "Meet" indicates that the software met the expectations of the scenario for the device in question. Since the scenarios are acquisition oriented, this ranking generally means that all of the identified data was successfully recovered. One caveat is that some phones lack the capability to handle certain data prescribed under a scenario, in which case the ranking applies only to the relevant subset. Similarly, the entry "Below" indicates that the software fell short of fully meeting expectations, while "Above" indicates that the software surpassed expectations.

A "Below" ranking is often a consequence of a tool performing a logical acquisition and being unable to recover deleted data, which is understandable. However, the ranking may also be due to active data placed on the device not being successfully recovered, which is more of a concern. An "Above" ranking is typically a result associated with the characteristics of a device, such as the reset function not completely deleting data and leaving remnants for recovery by the tool. "Above" rankings should only occur with the last two phone scenarios: Cleared Devices and Power Loss. The entry "Miss" indicates that the software unsuccessfully met any expectations, highlighting a potential area for improvement. Finally, the entry "NA" indicates that a particular scenario was not applicable to the device.

Synopsis of Device Seizure

Device Seizure version 1.1[5] is able to acquire information from Pocket PC, Palm OS, and BlackBerry devices, including those with cellular capabilities, SIM cards and both GSM and non-GSM cell phones. Device Seizure allows the examiner to connect a device via a USB or a Serial connection. Examiners must have the correct cables and cradles to ensure connectivity, compatible synchronization software, and a backup battery source available. Synchronization software (e.g., Microsoft ActiveSync, Palm HotSync, BlackBerry desktop manager software) allows examiners to create a guest partnership between the forensic workstation and the device under investigation.

Pocket PC Phone

Device Seizure acquires a Pocket PC Mobile phone device is done through Device Seizure with the aid of Microsoft's ActiveSync communication protocol. An examiner creates an ActiveSync connection as a "Guest" to the device. The "Guest" account is essential for disallowing any content synchronization between the workstation and the device prior to acquisition. Before acquisition begins, Device Seizure places a small `dll` program file on the device in the first available block of memory, which is then removed at the end of acquisition. Paraben indicated that Device Seizure uses the `dll` to access unallocated regions of memory on the device.

To get the remaining information, Device Seizure utilizes Remote API (RAPI)[6], which provides a set of functions for desktop applications to communicate with and access information on Windows CE platforms. These functions are accessible once a Windows CE device is connected through ActiveSync. RAPI functions are available for the following:
- Device system information – includes version, memory (total, used, and available), and power status retrieval
- File and directory management – allows retrieval of path information, find specific files, permissions, time of creation, etc.
- Property database access – allows information to be gleaned from database information present on the device
- Registry manipulation – allows the registry to be queried (i.e., keys and associated value)

If the device is password protected, the correct password must be supplied before the acquisition stage begins, as illustrated below in Figure 4. If the correct password is not known or provided, connectivity cannot be established and the contents of the device cannot be acquired.

---

[5] Additional information on Paraben products can be found at: **http://www.paraben-forensics.com**
[6] Additional information on RAPI can be found at: **http://www.cegadgets.com/artcerapi.htm**

Figure 4: Password Prompt (Pocket PC)

During the beginning stages of acquisition, the examiner is prompted with four choices of data to acquire as illustrated below.

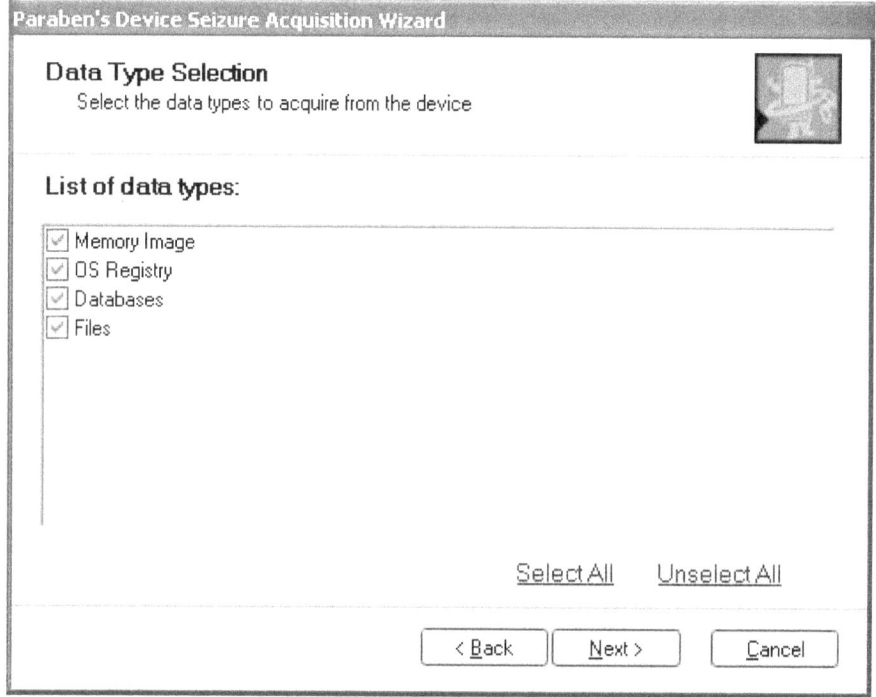

Figure 5: Acquisition Selection (Pocket PC)

Palm OS Phone

The acquisition of a Palm OS device with cell phone capabilities entails the forensic examiner exiting all active HotSync applications and placing the device in console mode. Console mode is used for physical acquisition of the device.[7] To put the Palm OS device in console mode, the examiner must go to the search window (press the magnifying glass by the Graffiti writing area), enter via the Graffiti interface the following symbols: lower-case cursive L, followed by two dots (results in a period), followed by writing a "2" in the number area. For acquiring data from a palmOne Treo 600, the technique used is slightly different. Instead of entering console mode via

---

[7] Additional information on console mode can be found at: http://www.ee.ryerson.ca/~elf/visor/dot-shortcuts.html

the Graffiti writing area, the shortcut used must be entered via the QWERTY keyboard. Console mode is device-specific and the correct sequence of graffiti characters can be found at the manufacturer's Web site. If the device is password protected, the proper password must be entered before acquisition. During the beginning stages of acquisition the examiner is prompted with the two choices of data elements to acquire as illustrated in Figure 6.

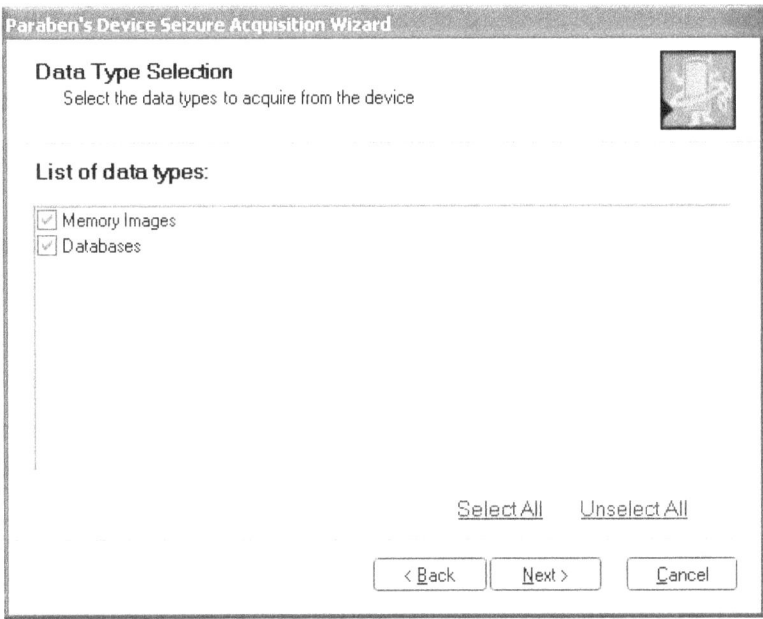

Figure 6: Acquisition Selection (Palm OS)

BlackBerry Device

The acquisition of a BlackBerry device is done through Device Seizure without the aid of synchronization protocols or the BlackBerry Desktop Manager. The BlackBerry Desktop Manager does allow users to upload applications, perform backups and restorations, and synchronization of defined data (e.g., Address Book, Calendar, Memo Pad, Tasks). Figure 7 shows a dialog box presented to the examiner before acquisition begins, if the BlackBerry device is password protected. If the password is not correctly supplied within 10 attempts all data is lost from the device and the BlackBerry OS has to be reinstalled.

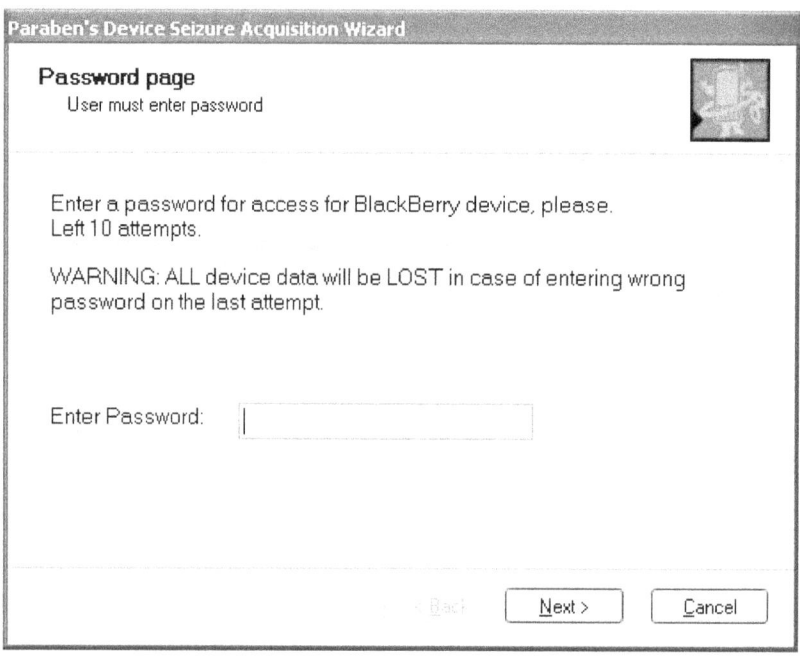

Figure 7: Password Prompt (BlackBerry)

After device selection the examiner is prompted with the following options shown in Figure 8 of acquiring either individual databases, memory, or both.

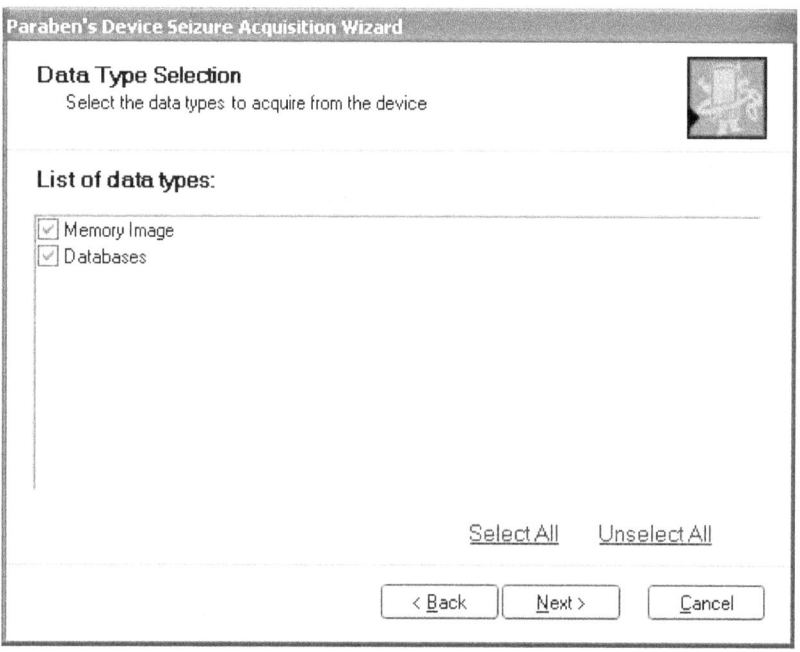

Figure 8: Acquisition Selection (BlackBerry)

If Acquire Databases and Memory are both selected, the memory is acquired first and then the individual databases are acquired. Before acquisition begins, the examiner is given the choice to pause between memory and database acquisitions as illustrated in Figure 9.

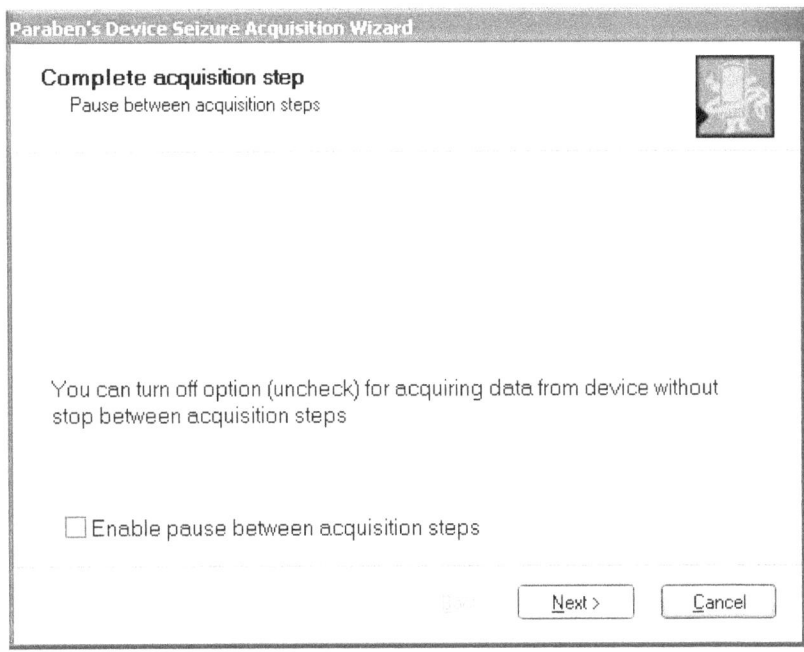

Figure 9: BlackBerry Acquisition

Cell Phone

Device Seizure version 1.1 is able to acquire information from various manufacturers' cell phones, including Motorola, Nokia, Samsung, Siemens, and Sony Ericsson.[8] The make, model, and type of phone determine how much data, if any, that Device Seizure can acquire. Typically, the following data can be acquired from the majority of cellular devices: Phone Calls Made, Phone Calls Received, Text Messages, and Phonebooks. Additionally, Device Seizure can acquire the following on some supported models: To-Do Lists, Calendar, Call Times, Call Info, Call History, Contact info, Phone Number of Cell, Image Gallery (e.g., Wallpapers, Camera Phone Images), Ring Tones, Reminders, Memos, Voice Memos, Events, Profiles, Games, Logos, WAP Settings, WAP Bookmarks, GPRS Access Points, Java files, and a complete Memory Dump. The breadth and depth of information acquired depends on the make, model, and network on which the phone operates. Device Seizure also allows independent acquisition of SIM cards with the included SIM card reader that comes with the purchase of Paraben's Toolbox. The toolbox provides all the necessary cables to create connectivity between supported phone models and the forensic workstation.

Acquisition Start

Two methods exist to begin the acquisition of data from the device or SIM. The acquisition can be enacted through the toolbar using the Acquire icon or through the Tools menu and selecting Acquire Image. Either option starts the acquisition process. With the acquisition process, both files and memory images can be acquired. By default, the tool marks both types of data to be acquired. Once the acquisition process is selected, the acquisition wizard illustrated below in Figure 10 guides the examiner through the process.

---

[8] Additional information on supported phone models can be found at: **http://www.paraben-forensics.com/cell_models.html**

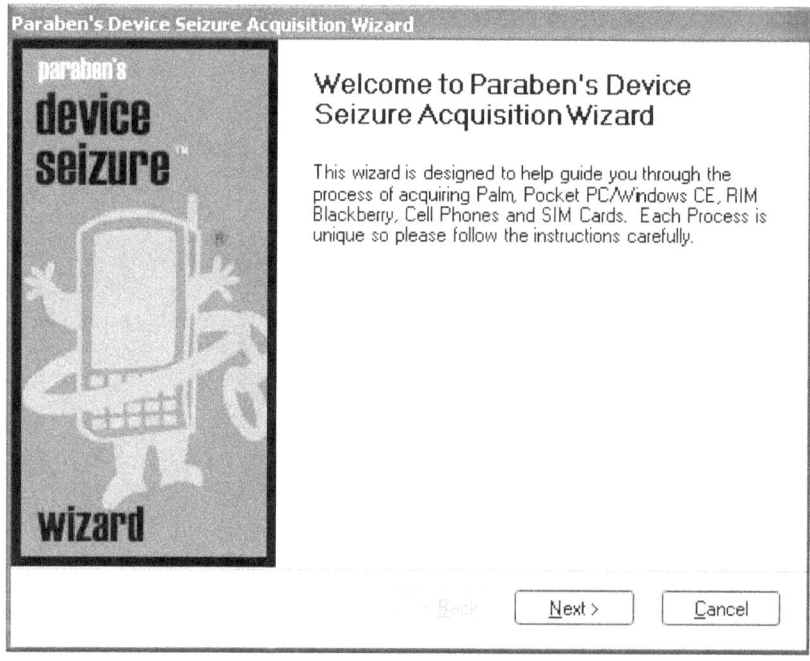

Figure 10: Acquisition Wizard

After clicking next on the Acquisition Wizard, the examiner is prompted to select which type of device to acquire, as illustrated below.

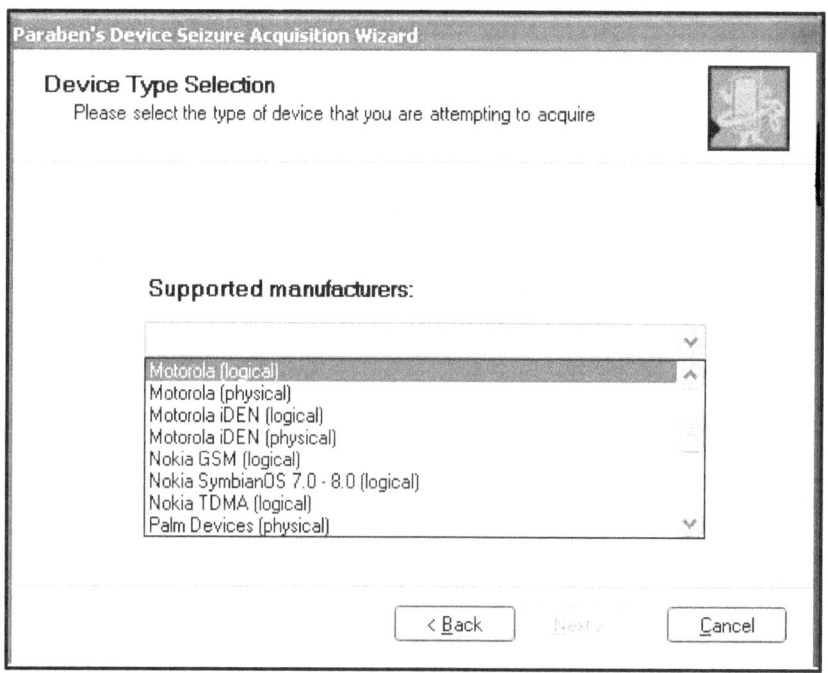

Figure 11: Acquisition Selection

Figure 12 below contains an example screen shot of Device Seizure during the acquisition of a Pocket PC device, displaying the various fields provided by the interface.

33

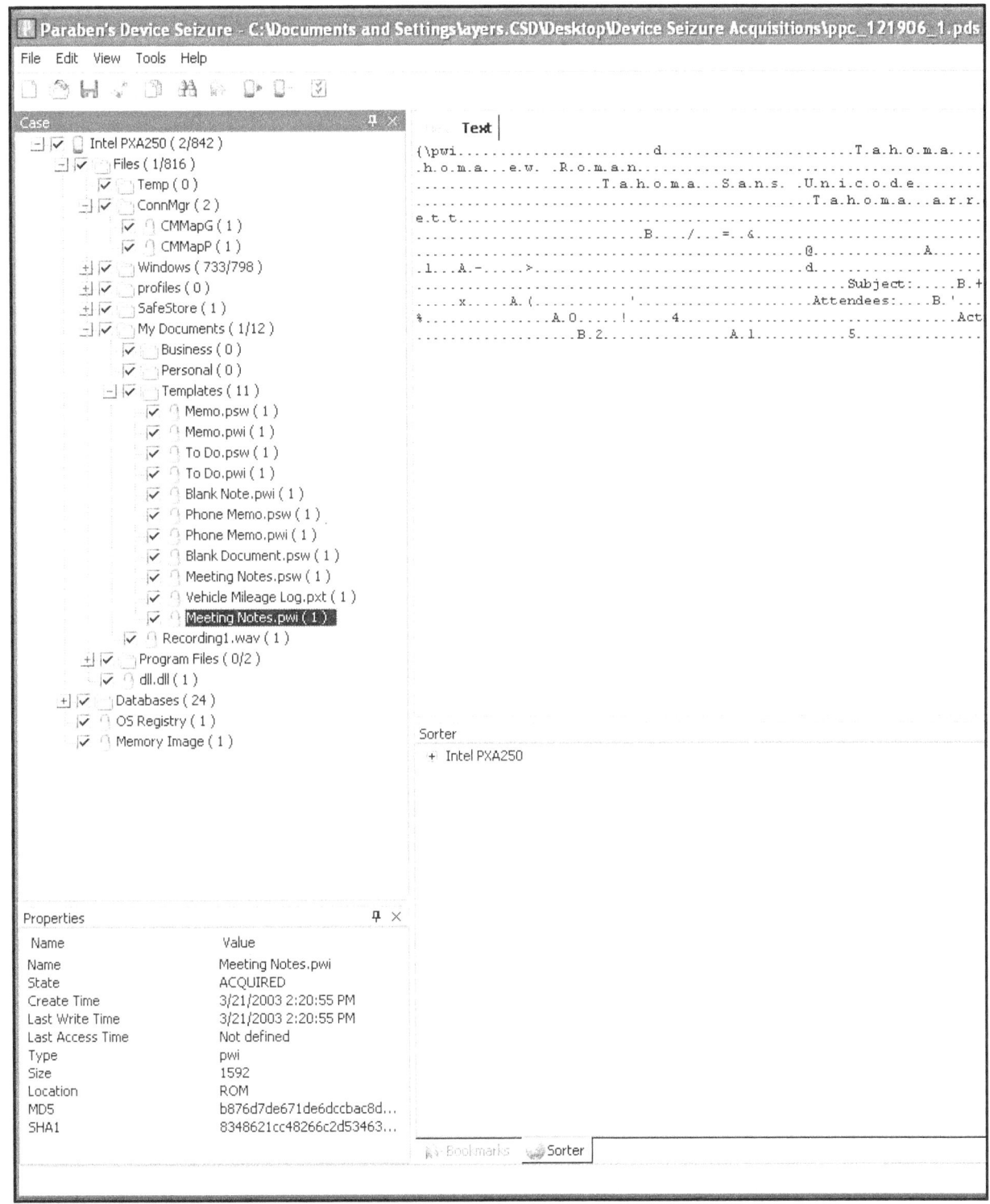

Figure 12: Acquisition Screen Shot (Pocket PC)

Device Seizure reports the following for each individual file acquired: File Path, File Name, File Type, Creation and Modification Dates, File Attributes, File Size, Status and an MD5 File Hash. Validation of file hashes taken before and after acquisition can be used to detect whether files have been modified during the acquisition stage.

Search Functional

Device Seizure's search facility allows examiners to query files for content. The search function searches the content of files and reports all instances of a given string found. Wildcards and regular expression are supported providing examiners with the ability to narrow down the number of false-positive returns. The screen shot shown below in Figure 13 illustrates an example of the results produced for the string "h.o.m.e.r".

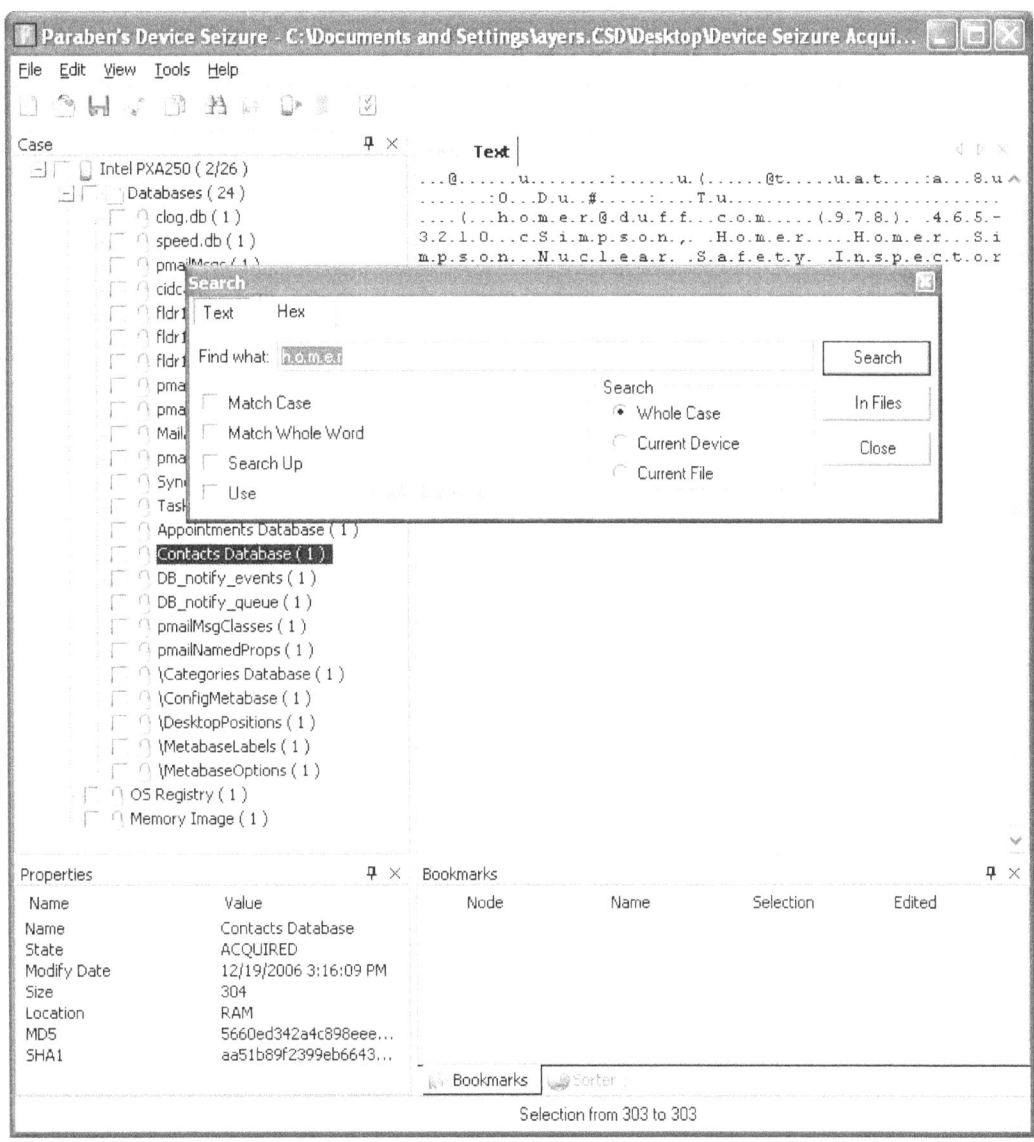

Figure 13: File Content String Search (Pocket PC)

Additionally the search window provides an output of memory related to the string search provided by the examiner. This allows examiners to scroll through sections of memory and bookmark valuable information for reporting to be used in judicial, disciplinary, or other proceedings.

Graphics Library

The graphics library enables examiners to examine the collection of graphics files present on the device. Deleted graphics files do not appear in the library. In order to display all graphics files present on the device, examiners must select the "Fill Sorter" function located beneath the "Tools" menu. The "Sorter" selects graphics files based on file signature not file extension. If deleted graphics files exist, they must be identified via the memory window by performing a string search to identify file remnants. However, recovery of an entire image is difficult, since its contents may be compressed by the filesystem or may not reside in contiguous memory locations, or may have some unrecoverable sections. Recovery also requires knowledge of associated data structures to piece the parts together successfully. Figure 14 shows a screen shot of images acquired from a Nokia 6610i device.

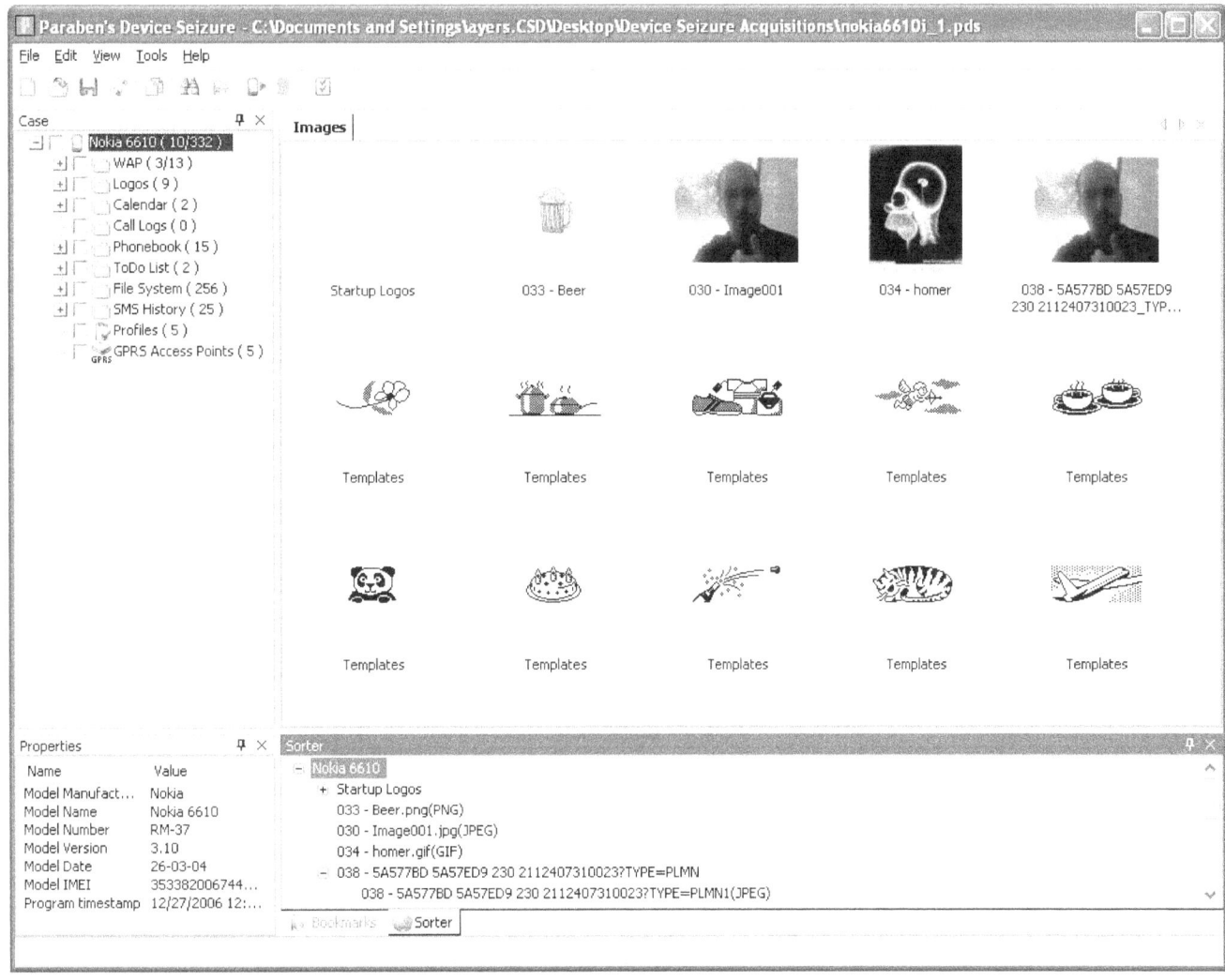

Figure 14: Graphics Library

Bookmarking

During an investigation, forensic examiners often have an idea of the type of information for which they are looking, based upon the circumstances of the incident and information already obtained. Bookmarking allows forensic examiners to mark items that are relevant to the investigation. Such a capability gives the examiner the means to generate case-specific reports containing significant information found during the examination, in a format suitable for presentation. Bookmarks can be added for multiple pieces of information found and each individual file can be exported for further analysis if necessary. Illustrated in Figure 15 below is an example of various bookmarks created after acquisition. As mentioned earlier, the files found and bookmarked can be exported to the workstation and rendered with an application suitable for the type of file in question.

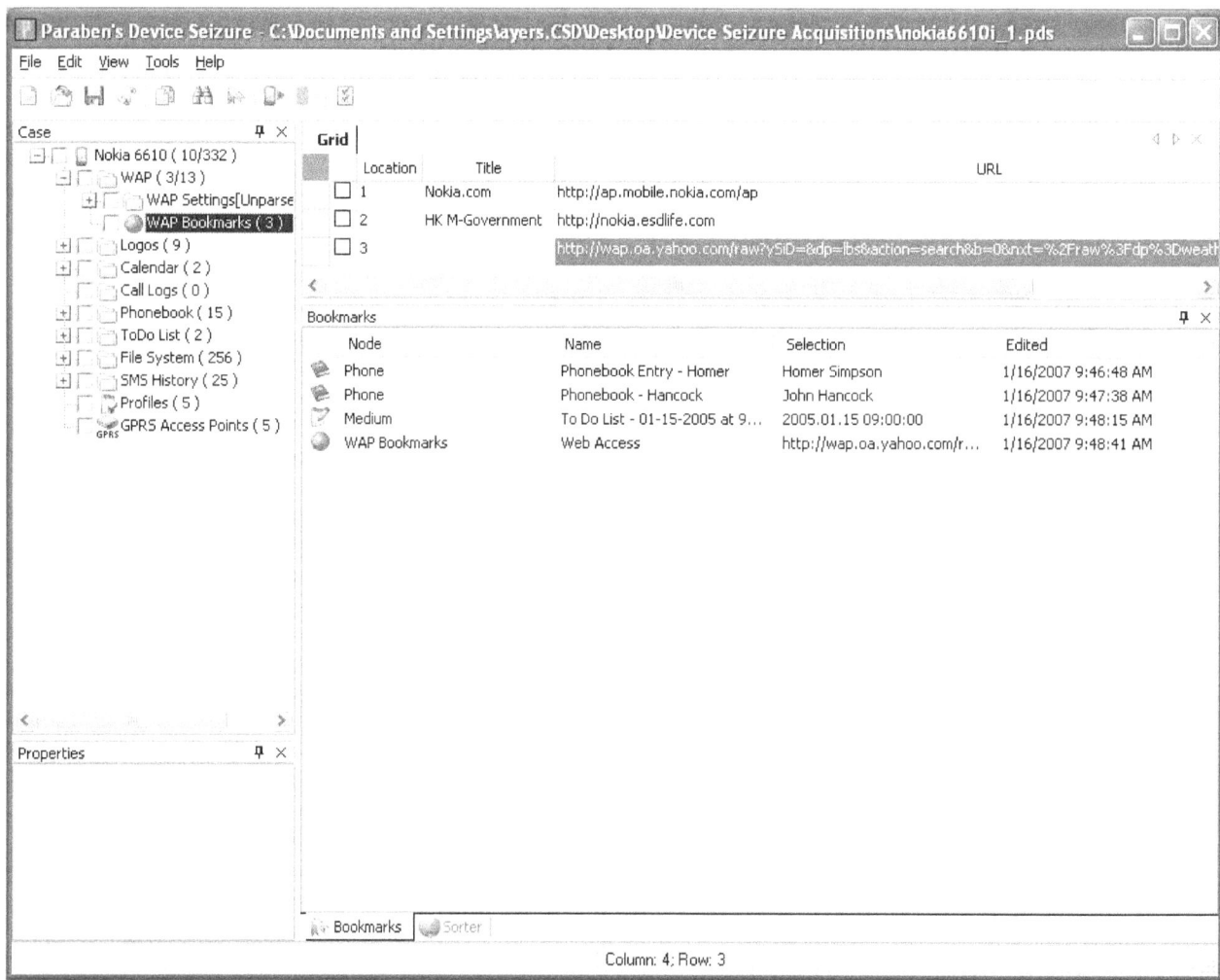

Figure 15: Bookmark Creation (Nokia 6610i)

Additional Tool

Export All Files: Examiners can export all files reported after the acquisition stage has been completed. When the files are exported, a folder is created, based upon the case file name, containing two subfolders: one each for RAM and ROM. Depending upon the type of file, the contents can be viewed with an associated desktop application or with a device specific emulator.

Validate Hash Codes: The Validate Hash Codes option can be run after a successful acquisition and is designed to report files that have been altered during the acquisition process.

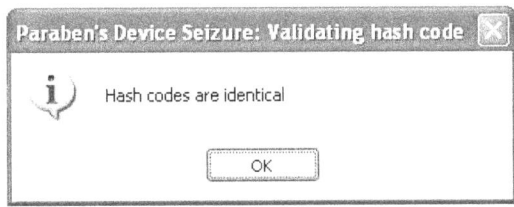

Figure 16: Validate Hash Codes

Case Comparer: Device Seizure has a built-in function that compares acquisition case-files. To operate the compare feature, case-files in question are imported via the Tools menu and automatically scanned for differences. The results are shown in a dialog box listing the file name, the result of the compare, and the size of each case (.pds) file. Double-clicking a file in question provides a side-by-side hex view of the two files with the differences high-lighted. Device Seizures Case Comparer is illustrated below in Figure 17.

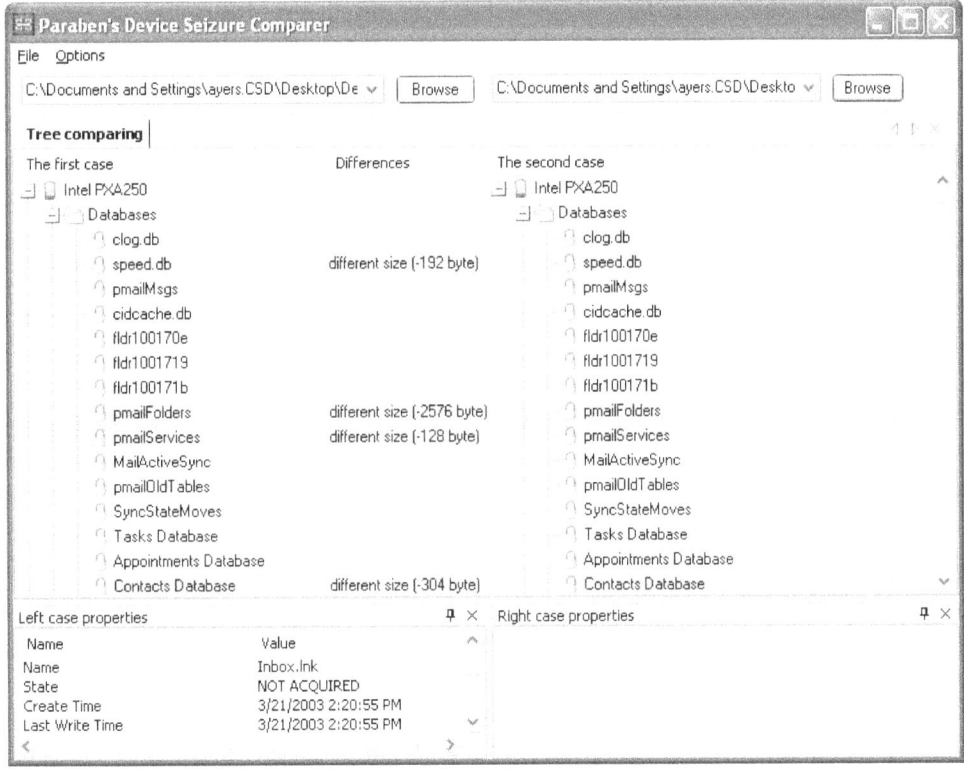

Figure 17: Case Comparer

Device Seizure File Viewer: Each file acquired can be viewed in either text or hex mode, allowing examiners to inspect the contents of all files present. Examiners must use one of the following options to rendition other types of files: export the file and launch a Windows application corresponding to the file extension (Run File's Application), or, for Palm OS devices only, view the file through the POSE.

Report Generation

Device Seizure provides a user interface for report generation that allows examiners the entry and organization of case-specific information. Each case contains an identification number and other information specific to the investigation for reporting purposes, as illustrated in Figure 18 below.

Once the report has been generated, it produces an .html file is available for the examiner, which includes bookmarked files, total files acquired, acquisition time, device information. If files were modified during the acquisition stage, the report identifies them.

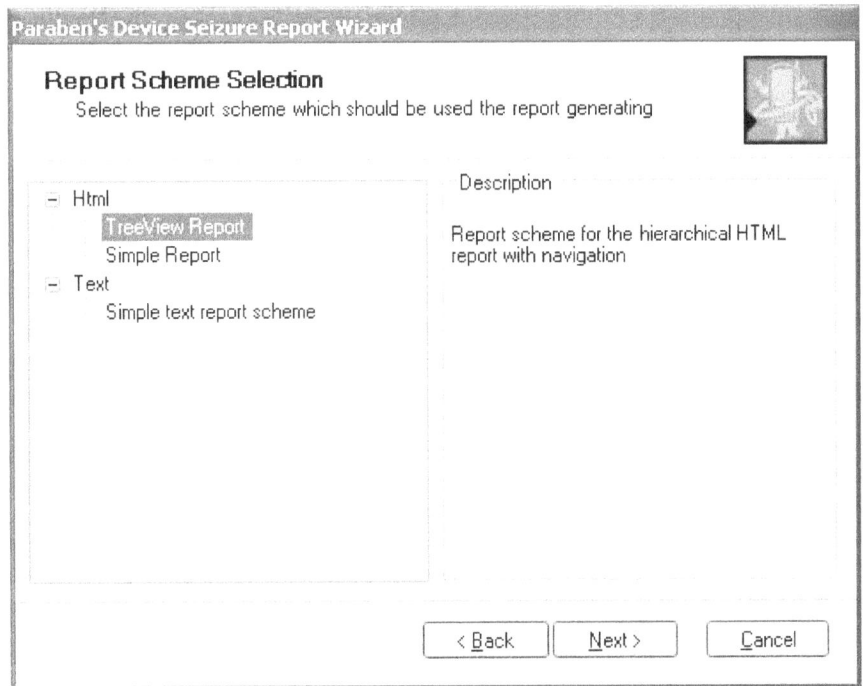

Figure 18: Report Generation

Scenario Results - PD

Table 9 summarizes the results from applying the scenarios listed at the left of the table to the devices across the top. More information can be found in Appendix A: Device Seizure Results.

Table 9: Results Matrix

Scenario	Device					
	BlackBerry 7750	BlackBerry 7780	Kyocera 7135	Motorola MPX220	Samsung i700	Treo 600
Connectivity and Retrieval	Meet	Meet	Meet	Meet		Meet
PIM Applications	Meet	Meet	Meet	Miss	Meet	Meet
Dialed/Received Phone Calls	Meet	Meet	Meet	Miss	Meet	Meet
SMS/MMS Messaging	Meet	Meet	Below	Below	Below	Meet
Internet Messaging	Meet	Meet	Below	Below	Below	Below

Scenario	Device					
	BlackBerry 7750	BlackBerry 7780	Kyocera 7135	Motorola MPX220	Samsung i700	Treo 600
Web Applications	NA	Below	Below	Below	Below	Below
Text File Formats	Below	Below	Meet	Below	Below	Meet
Graphics File Formats	Miss	Miss	Miss	Below	Below	Miss
Compressed Archive File Formats	Miss	Miss	Miss		Meet	Meet	Below
Misnamed Files	Miss	Miss	Meet	Meet	Meet	Meet
Peripheral Memory Cards	NA	NA	Miss	Below	Below	Miss
Acquisition Consistency	Meet	Meet	Below	Meet	Below	Below
Cleared Devices	Meet/Above	Meet/Above	Meet	Meet	Meet	Meet
Power Loss	Above	Above	Meet	Above	Meet	Meet

Scenario Results - Cell Phones

Table 10 summarizes the results from applying the scenarios listed at the left of the table to the devices across the top. More information can be found in Appendix B: Device Seizure Results.

Table 10: Results Matrix

Scenario	Devices								
	Audiovox 8910	Ericsson T68i	LG 4015	Motorola C333	Motorola v66	Motorola v300	Nokia 3390	Nokia 6610i	Sanyo PM8200
Connectivity and Retrieval	Meet	Meet	Meet	Meet	Meet	Meet	Meet	Meet	
PIM Applications	Meet	Below	Below	Meet	Below	Below	Below	Below	Meet
Dialed/Received Phone Calls	Meet	Below	Miss	Below	Below	Below	Below	Below	Meet
SMS/MMS Messaging	Below	Below	Below	Meet	Below	Meet	Below	Below	Meet
Internet Messaging	NA	Miss	NA	NA	NA	Meet	NA	NA	Miss
Web Applications	Below	Miss	Miss	NA	NA	Miss	Miss	Below	Miss
Text File Formats	NA	NA	NA	NA	NA	Miss	NA	Below	NA
Graphics File Formats	Below	Miss	NA	NA	NA	Meet	NA	Below	Miss
Compressed Archive File Formats	NA	NA	NA	NA	NA	NA	Meet	NA	
Misnamed Files	NA	NA	NA	NA	NA	Miss	NA	Meet	NA
Peripheral Memory Cards	NA	NA	NA	NA	NA	NA	NA	NA	NA
Acquisition Consistency	Meet	Meet	Meet	Meet	Meet	Meet	Meet		Meet
Cleared Devices	NA	Meet	Above	Above	Above	Above	NA	NA	Above
Power Loss	Above	Above	Above	Above	Above	Above	Above	Above	Above

Scenario Results - SIM Card Acquisition

Device Seizure allows examiners the ability to acquire data directly from a SIM card with the use of Paraben's RS-232 SIM Card Reader. The acquisition steps followed to acquire data directly from the SIM are the same as acquiring data from a phone except for selecting GSM SIM Card. The data fields acquired (e.g., Abbreviated Dialing Numbers, Fixed Dialing Numbers, Last Numbers Dialed, SIM Service Dialing Numbers, Short Messages) depend on the SIM and service provider. The Search Functionality, Bookmarking facilities, and Report Generation operate on the acquired data in a similar fashion to phone acquisitions, described above.

Table 11 summarizes the results from applying the scenarios listed at the left of the table to the SIMs across the top. More information can be found in Appendix G: SIM Seizure – External SIM Results. SIMs were acquired and reported using Paraben's SIM Seizure portion of Device Seizure.

Table 11: SIM Card Results Matrix - External Reader

Scenario	SIM		
	5343	8778	1144
Basic Data	Below	Below	Below
Location Data	Meet	Meet	Meet
EMS Data	Meet	Meet	Meet
Foreign Language Data	Meet	Meet	Meet

Synopsis of Pilot-link

The pilot-link[9] software can be used to obtain both the ROM and RAM from Palm OS devices (e.g., Palm, Handspring, Handera, TRGPro, Sony) over a serial connection. The data can be imported into the Palm OS Emulator (POSE), allowing a virtual view of the data contained on the device, or the individual files can be viewed with a standard ASCII hex editor or through a compatible forensic application. Additionally, the data created from pilot-link can be imported into other compatible forensic applications. Once the software is installed and configured, communications between the workstation and the device can begin. Pilot-link uses the HotSync protocol to acquire data from the device. RAM and ROM are dumped from the device with the following commands: pi-getrom and pi-getram. Individual database files can be acquired with the pilot-xfer –b command. Pilot-link lacks an integrated search engine, report generation facilities, and a graphics library. Therefore, third party tools must be utilized to perform these functions. Table 12 summarizes the results from applying the scenarios listed at the left of the table to the devices across the top. Additional information can be found in *NISTIR 7250 Cell Phone Forensic Tools: An Overview and Analysis*.

Table 12: Results Matrix

Scenario	Device	
	Kyocera 7135	Treo 600
Connectivity and Retrieval	Meet	Below
PIM Applications	Meet	Meet
Dialed/Received Phone Calls	Meet	Below
SMS/MMS Messaging	Meet	Below
Internet Messaging	Meet	Below
Web Applications	Below	Below
Text File Formats	Meet	Meet
Graphics File Formats	Below	Below
Compressed Archive File Formats	Miss	Miss
Misnamed Files	Meet	Meet
Peripheral Memory Cards	Miss	Miss
Acquisition Consistency	Below	Below
Cleared Devices	Meet	Meet
Power Loss	Meet	Meet

---

[9] Additional information on pilot-link can be found at: **http://www.pilot-link.org**

Synopsis of GSM .XRY

Micro Systemation's GSM .XRY version 3.0 can acquire information from various manufacturers of GSM cell phones (i.e., Ericsson, Motorola, Nokia, Siemens).[10] The .XRY unit provides the ability to perform acquisitions via cable, IrDA (Infrared), or Bluetooth interfaces. The GSM .XRY unit provides all the necessary cables to create connectivity between supported phone models and the forensic workstation.

Supported Phone

The make, model, and type of phone determine what data GSM .XRY can acquire. Micro Systemation's Web site provides a link to a soft copy version of the manual listing the supported make and models of cell phones. GSM .XRY is targeted at GSM and CDMA devices including 3G devices. TDMA phones are not currently supported. GSM .XRY can acquire the following information from GSM (Global System for Mobile Communications) phones: Contacts, Calls, Calendar, SMS, Pictures, Audio, Files, Notes, MMS, Video, Network Information and Tasks. Each type of data acquired appears in the menu bar.

Acquisition Stage

Two methods exist to acquire data from cell phones. The acquisition can be enacted through the toolbar, using the Extract data icon, or through the File menu, selecting Extract data from the device. Either option starts the acquisition process. With the acquisition process, both internal phone memory and basic SIM card information (e.g., phone book entries, SMS messages) are acquired. Once the acquisition process is selected, the acquisition wizard illustrated below in Figure 19 appears to guide the examiner through the process.

---

[10] Additional information on supported phone models can be found at: **www.msab.com**

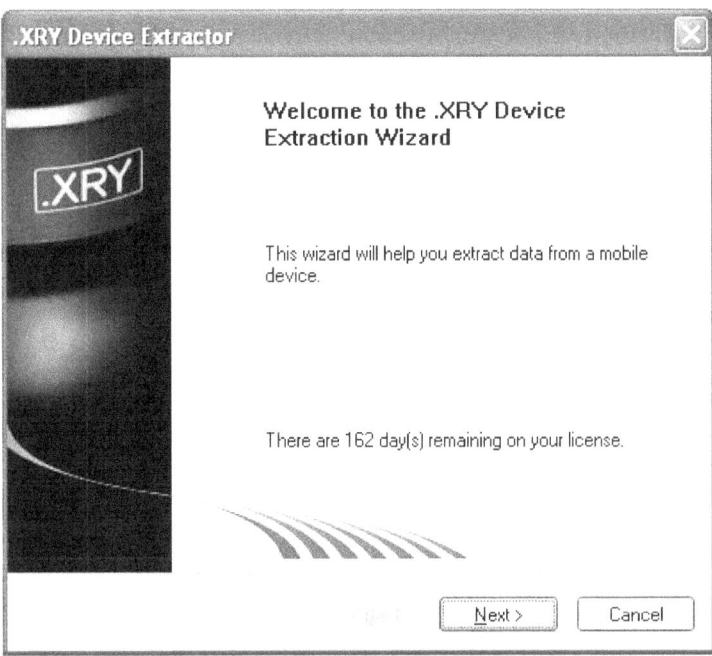

Figure 19: Acquisition Wizard

Following the execution of the acquisition wizard, the examiner selects the acquisition type: mobile device or SIM/USIM card. Assuming selection of a mobile phone device, the examiner is presented with three interface choices to create a connection with the cell phone. Micro Systemation provides specific recommendations (i.e., Cable, Infrared, or Bluetooth) for each make and model of phones supported.

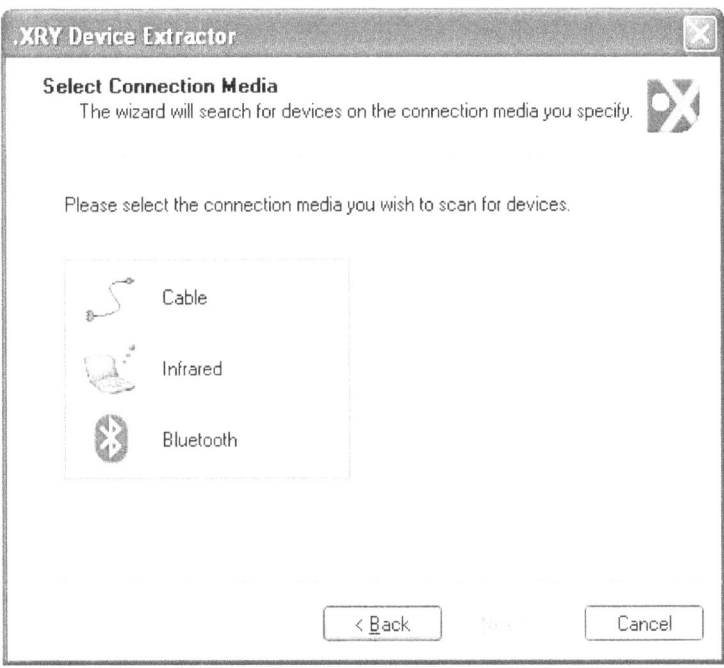

Figure 20: Interface Selection

After a successful connection has been established, the device is identified and the actual acquisition begins as illustrated below in Figure 21.

Figure 21: Device Identification

During the acquisition stage GSM .XRY keeps a process log of the status of information extracted from the device, as illustrated in Figure 22.

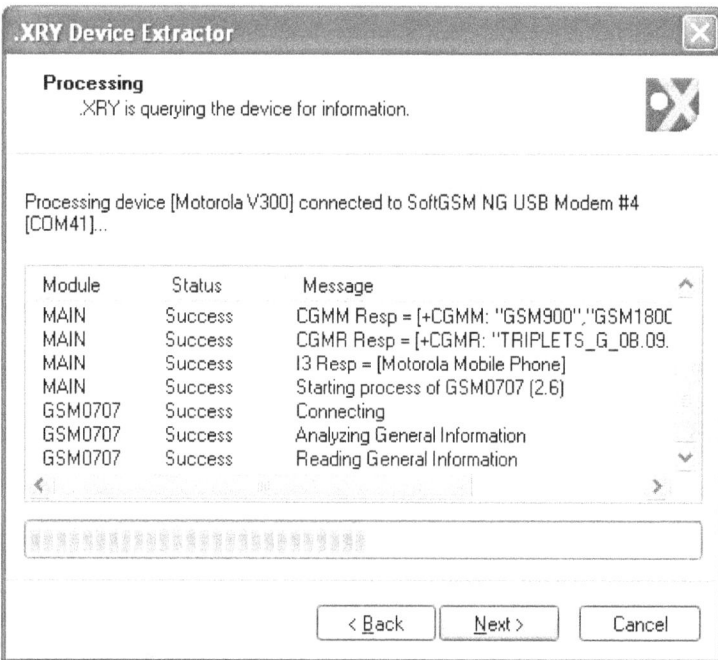

Figure 22: Acquisition Log

Search Functional

GSM .XRY's search facility allows examiners to query the acquired data for content. The search function scans the content of files and reports all instances of a given string found. The screen shot shown in Figure 23 illustrates an example of the search window options and results produced for the string "SMS message". Search hits that are found are highlighted in pink.

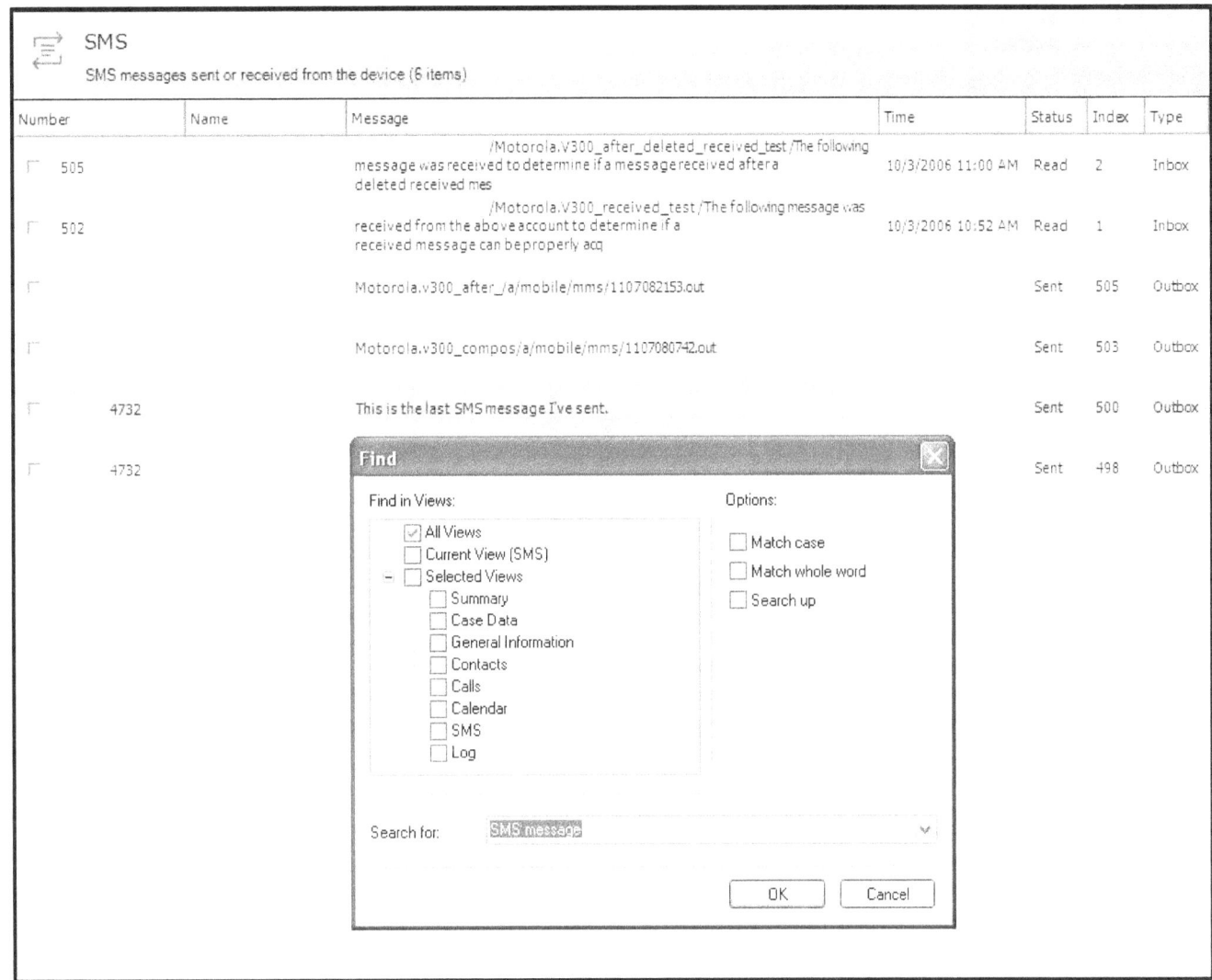

Figure 23: File Content String Search

Graphics Library

The graphics library enables examiners to examine the collection of graphics files present on the device. Each image present can be viewed internally with the Picture Window application, allowing examiners to enlarge images if necessary. Additionally, images collected can be exported and inspected with a third party tool, if necessary. Figure 24 shows a screen shot of images acquired from a Nokia 6610i.

Figure 24: Graphics Library

Report Generation

GSM .XRY allows customized reports to be created with predefined data selection, as illustrated in Figure 25. Bookmarking facilities do not exist in GSM .XRY. Therefore examiners cannot filter data within selected categories. Additionally, reports by default, do not embed an illustrative view of acquired graphics files; only filenames, file size, and meta-data are included. Graphical data is included in a separate folder when the report is exported.

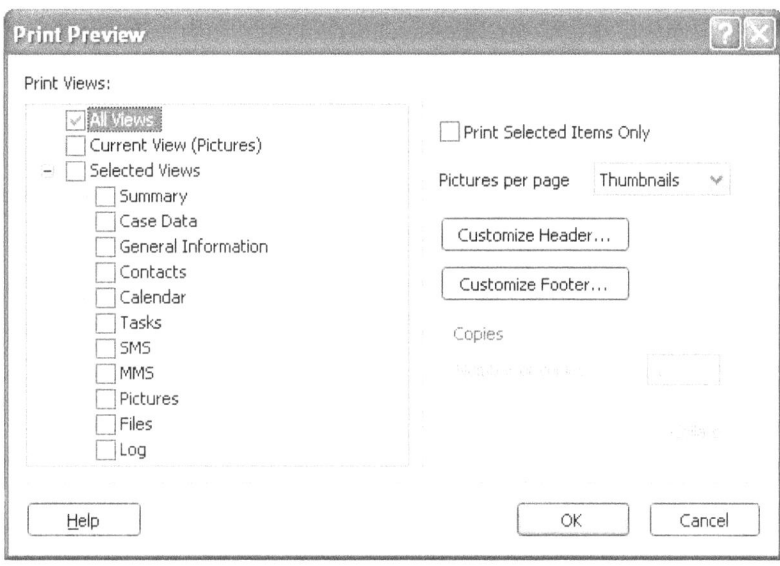

Figure 25: Report Generation

As stated above, examiners have the ability to include all data acquired from the cell phone or to choose a particular category of information. Illustrated below in Figure 26 and Figure 27 are snapshots of a report generated when choosing respectively the General Information view and a Pictures view.

Figure 26: Report Excerpt (General Information)

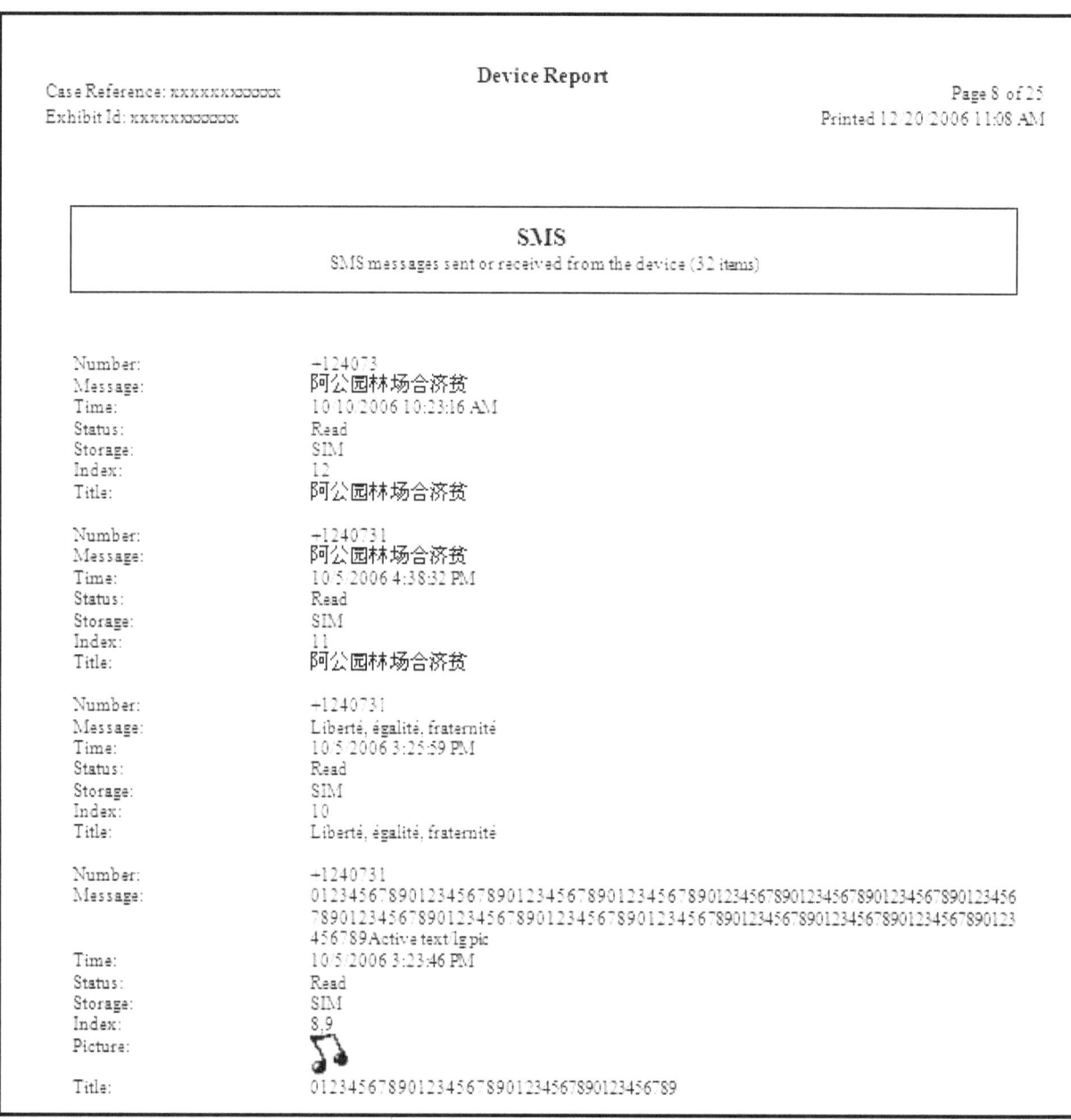

Figure 27: Report Excerpt (Text Messages)

Scenario Results – Cell Phones

Table 13 summarizes the results from applying the scenarios listed at the left of the table to the devices across the top. More information can be found in Appendix C: GSM .XRY Results.

Table 13: Results Matrix

Scenario	Device					
	Ericsson T68i	Motorola V66	Motorola V300	Nokia 6610i	Nokia 6200	Nokia 7610
Connectivity and Retrieval	Meet	Meet	Meet	Meet	Meet	
PIM Applications	Below	Below	Below	Below	Below	Below

Scenario	Device					
	Ericsson T68i	Motorola V66	Motorola V300	Nokia 6610i	Nokia 6200	Nokia 7610
Dialed/Received Phone Calls	Below	Below	Below	Below	Below	Miss
SMS/MMS Messaging	Below	Below	Below	Below	Below	Miss
Internet Messaging	Miss	NA	Below	NA	NA	Miss
Web Applications	Miss	NA	Miss	Below	Miss	Miss
Text File Formats	NA	NA	Miss	Below	Below	Below
Graphics File Formats	Miss	NA	Miss	Below	Below	Below
Compressed Archive File Formats	NA	NA	NA	Meet	Meet	Meet
Misnamed Files	NA	NA	Miss	Meet	Meet	Meet
Peripheral Memory Cards	NA	NA	NA	NA	NA	Below
Acquisition Consistency	NA	NA	NA	NA	NA	NA
Cleared Devices	Meet	Meet	Above	NA	NA	Meet
Power Loss	Above	Above	Above	Above	Above	Above

Scenario Results - SIM Card Acquis

GSM .XRY version 2.4 through 2.5 provide examiners the ability to acquire data directly from a SIM card using the CardMan reader by OMNIKEY. The acquisition steps followed to acquire data directly from the SIM are the same as acquiring data from a phone, except for selecting SIM/USIM Card from the user interface in lieu of Mobile Phone. The data fields acquired (e.g., General Information (i.e., ICCID, IMSI, Phase), Contacts, Calls, Messages) are dependent upon the SIM and service provider. The Search engine and Report facilities operate in a similar fashion as with phone acquisitions, described above.

Table 14 summarizes the results from applying the scenarios listed at the left of the table to the SIMs across the top. More information can be found in Appendix H: GSM .XRY – External SIM Results.

Table 14: SIM Card Results Matrix - External Reader

Scenario	SIM		
	5343	8778	1144
Basic Data	Meet	Meet	Meet
Location Data	Below	Below	Below
EMS Data	Meet	Meet	Meet
Foreign Language Data	Meet	Meet	Meet

Synopsis of Oxygen Phone Manager

Oxygen Phone Manager (OPM)[11] is a tool designed to manage information on a cell phone, including contacts, calendar, SMS messages, To-Do list, logs, and ring tones. The software is designed to support most Nokia phones. A Symbian OS version also is available. Oxygen Phone Manager - Forensic Version is an adaptation of the phone management tool that suppresses changing data on the phone, but allows data to be logically acquired and exported into several supported formats. The tool is designed to acquire phonebook contacts (including pictures), Call Lists (i.e., last numbers dialed, missed and received calls), SMS messages, pictures, logos, ring tones, profiles, To-Do lists, MMS messages (supported formats are plain text, HTML, JPEG, GIF, animated GIF, PNG, TIFF, BMP, MIDI, WAV, and RT), Java applications, games, gallery and play lists. OPM does not provide search capabilities. However, a tree structure of the acquired data is built and populated with the selected data, providing examiners with the ability to browse acquired files and view graphical images. All the items can be saved or exported in common formats and searched using a third party tool. Internal report generation is not provided. The data extracted from the phone through the tool interface must be exported manually and processed through another means, such as a word processor. Data can be saved in several different formats (e.g., .rtf, .html, .csv, etc.) for export. Certain data, such as Profiles, can only be saved in the OPM proprietary data format. Table 15 summarizes the results from applying the scenarios listed at the left of the table to the devices across the top. Additional information can be found in *NISTIR 7250 Cell Phone Forensic Tools: An Overview and Analysis*.

Table 15: Results Matrix

Scenario	Device			
	Nokia 3390	Nokia 6610i	Nokia 6200	Nokia 7610
Connectivity and Retrieval	Meet	Meet	Meet	Meet
PIM Applications	Below	Below	Below	Below
Dialed/Received Phone Calls	Below	Below	Below	Miss
SMS/MMS Messaging	Below	Below	Below	Below
Internet Messaging	NA	NA	NA	Below
Web Applications	Miss	Miss	Miss	Miss
Text File Formats	NA	Below	Below	Miss
Graphics File Formats	NA	Below	Below	Below
Compressed Archive File Formats	NA	Meet	Meet	Miss
Misnamed Files	NA	Meet	Meet	Miss
Peripheral Memory Cards	NA	NA	NA	Below
Acquisition Consistency	NA	NA	NA	NA
Cleared Devices	NA	NA	NA	Meet
Power Loss	Above	Above	Above	Above

---

[11] Additional information can be found at: **http://www.opm-2.com/forensic**

Synopsis of MOBILedit!

MOBILedit! Forensic, version 2.0.0.10, is able to acquire information from various GSM, CDMA, and PCS cell phones from assorted manufacturers (i.e., Alcatel, Ericsson, General, LG, Motorola, Nokia, Panasonic, Philips, Samsung, Siemens, Sony Ericsson).[12] A non-forensic variant of the product, called MOBILedit!, also exists. This report covers only the forensic version, which for simplicity is referred to as MOBILedit!. The MOBILedit! application provides the ability to perform acquisitions via cable, IrDA (Infrared), or Bluetooth interfaces.

The information acquired by MOBILedit! depends on the make, model, and richness of phone features. Some common data fields acquired using MOBILedit! are phone and subscriber information, Phonebook, SIM Phonebook, Missed calls, Last Numbers Dialed, Received calls, Inbox, Sent items, Drafts, and Files (i.e., Graphics, Photos, Tones).

MOBILedit's search engine allows examiners to perform simplified string searches only within specific folders. The search engine does not provide the ability to search through multiple cases or multiple folders within a case, or issue complex expression patterns. The graphics library enables examiners to examine the collection of graphics files present on the device. Each image present can be viewed internally with the Picture Window application allowing examiners to enlarge images if necessary. Additionally, images collected can be exported and inspected with a third party tool, if necessary. Version 1.95 and above incorporate a report generation facility, allowing examiners to produce reports internally within the application or to export them in an xml format.

Table 16 summarizes the results from applying the scenarios listed at the left of the table to the devices across the top. Additional information can be found in *NISTIR 7250 Cell Phone Forensic Tools: An Overview and Analysis*.

Table 16: Results Matrix

Scenario	Device				
	Ericsson T68i	Motorola C333	Motorola V66	Motorola V300	Nokia 6610i
Connectivity and Retrieval	Meet	Meet	Meet	Meet	Meet
PIM Applications	Miss	Below	Below	Below	Below
Dialed/Received Phone Calls	Below	Below	Below	Below	Below
SMS/MMS Messaging	Below	Miss	Miss	Below	Below
Internet Messaging	Miss	NA	NA	Below	NA
Web Applications	Miss	NA	NA	Miss	Miss
Text File Formats	NA	NA	NA	Miss	Below
Graphics File Formats	Miss	NA	NA	Miss	Below

---

[12] Additional information on supported phone models can be found at: **www.mobiledit.com**

Scenario	Device				
	Ericsson T68i	Motorola C333	Motorola V66	Motorola V300	Nokia 6610i
Compressed Archive File Formats	NA	NA	NA	NA	Meet
Misnamed Files	NA	NA	NA	NA	Meet
Peripheral Memory Cards	NA	NA	NA	NA	NA
Acquisition Consistency	NA	NA	NA	NA	NA
Cleared Devices	Meet	Above	Meet	Above	NA
Power Loss	Above	Above	Above	Above	Above

SIM Card Acquisition

Mobiledit! Forensic version 2.0.0.10 and above gives examiners the ability to acquire SIM card data using a PC/SC-compatible reader. The acquisition steps followed to acquire data directly from the SIM are the same as acquiring data from a phone, except for selecting Smart Card Readers instead of Mobile Phones. The data fields acquired (i.e., SIM Phonebook, Last Numbers Dialed, Fixed Dialing Numbers, Inbox, Sent Items, Drafts) dependent in part on the SIM and service provider. The Search engine and Report facilities operate in a similar fashion as for phone acquisitions, described above.

Table 17 summarizes the results from applying the scenarios listed at the left of the table to the devices across the top. Additional information can be found in *NISTIR 7250 Cell Phone Forensic Tools: An Overview and Analysis*.

Table 17: SIM Card Results Matrix – External Reader

Scenario	SIM		
	5343	8778	1144
Basic Data	Below	Below	Below
Location Data	Miss	Miss	Miss
EMS Data	Below	Below	Below
Foreign Language Data	Below	Below	Below

Synopsis of BitPIM

BitPIM version 0.9.10 can acquire information primarily from various manufacturers of CDMA cell phones (e.g., Audiovox, Samsung, Sanyo).[13] The BitPIM application provides the ability to perform acquisitions through a cable interface. The make, model, and type of CDMA phone determine what data BitPIM can acquire. Some common data fields that BitPIM recovers are: Phonebook, Wallpapers (graphic files present on the phone), Ringers (sound bites), Calendar entries, text messages, call history, Memo entries and phone lock codes. BitPIM also captures a

---

[13] Additional information on supported phone models can be found at: **www.bitpim.org**

logical dump of the filesystem, where data related to Incoming/Outgoing/Missed/Attempted calls and SMS/MMS messages can be found. The filesytem data dump allows examination of SMS and MMS, message content and potential recovery of deleted items related to Phonebook entries and incoming and outgoing messages.

BitPIM provides no search functionality. However, items can be saved or exported in common formats and searched using a third party tool. The Media tab enables examiners to examine the collection of graphics files present on the device. Each image present can be viewed internally with the Picture Window application, which allows images to be enlarged. Additionally, collected images can be exported and inspected with a third party tool, if necessary. BitPIM does not support reporting facilities internally. Relevant data collected from the device can be copied by right clicking on a specified item and pasted elsewhere. Third party tools or editors can be used to create a finalized report of significant findings. Table 18 summarizes the results from applying the scenarios listed at the left of the table to the devices across the top. Additional information can be found in *NISTIR 7250 Cell Phone Forensic Tools: An Overview and Analysis*.

Table 18: Results Matrix

Scenario	Device	
	Audiovox 8910	Sanyo 8200
Connectivity and Retrieval	Meet	Meet
PIM Applications	Meet	Meet
Dialed/Received Phone Calls	Meet	Meet
SMS/MMS Messaging	Below	Meet
Internet Messaging	NA	Miss
Web Applications	Below	Below
Text File Formats	NA	NA
Graphics File Formats	Below	Below
Compressed Archive File Formats	NA	NA
Misnamed Files	NA	NA
Peripheral Memory Cards	NA	NA
Acquisition Consistency	NA	NA
Cleared Devices	NA	Above
Power Loss	Above	Above

Synopsis of TULP2G

TULP2G is a forensic software framework developed by the Netherlands Forensic Institute (NFI) for extraction and decoding of data stored in electronic devices.[14] The TULP2G

---

[14] Additional information on TULP2G can be found at: **http://tulp2g.sourceforge.net**

framework involves an abstract architecture, with distinct plug-in interfaces for data extraction through various means, data decoding of the extracted data, and also a user interface. TULP2G is not designed for presentation, viewing, or searching of extracted information; it uses XML for its data storage format and relies on existing tools for performing these functions. The framework, along with number of different data extraction and decoding plug-ins for cell phones, has been implemented as open source software.

Version 1.1.0.2 of TULP2G can acquire evidence from a phone through different means of communication (i.e., cable, Bluetooth, IrDA) and protocols (i.e., ETSI and Siemens AT commands, IrDA, and OBEX). For GSM phones, it can also acquire SIM data through an external PC/SC reader. The tool was designed to work for a wide variety of phones that support one or more common interface standards. Following this approach, the tool performs a logical acquisition using selected commands from the different protocol standards available for USB, IrDA, serial modem interfaces and PC/SC readers. Currently, a modem connection (i.e., serial port, serial over either USB, Bluetooth, or IrDA), a socket connection (i.e., IrDA, Bluetooth), and a PC/SC connection are supported. The appropriate protocol takes place over these connections (i.e., modem:AT_ETSI, AT_Siemens; IrDA: IrMC; Bluetooth: OBEX; PC/SC: SIM chip card data extraction). The data is then transformed using the corresponding conversion plug-ins (e.g., AT_ETSI, SMS) and an XML file is produced. A style sheet can be applied to the file to generate reports in a variety of formats and content.

TULP2G can acquire phone calls made, phone calls received, SMS messages, and phone book entries for various phones. It also can acquire more precise data (e.g., sent/received email or calendar, To-Do list), depending on the communication/protocol pair chosen. TULP2G also acquires IMEI, IMSI and specific data about the phone. Because TULP2G is not a search and analysis tool, after the XML or HTML formatted report is generated, examiners must manually search the report for data, or export the data and use some other search facility.

Graphic files are stored in the XML output file with <![CDATA[tags and can be converted and recovered using the conversion plug-ins (e.g., OBEX, SMS TPDU). Once the data has been converted properly, individual files can be saved to the desktop for reporting purposes. Table 19 summarizes the results from applying the scenarios listed at the left of the table to the devices across the top. Additional information can be found in *NISTIR 7250 Cell Phone Forensic Tools: An Overview and Analysis*.

Table 19: Results Matrix

Scenario	Device							
	Audiovox 8910	Ericsson T68i	Sony Ericsson P910a	Motorola C333	Motorola V66	Motorola V300	Nokia 6610i	Nokia 6200
Connectivity and Retrieval	Meet	Meet	Meet	Meet	Meet	Meet		Meet
PIM Applications	Below	Below	Miss	Below	Below	Below		Below
Dialed/Received Phone Calls	Below	Below	Miss	Below	Below	Below		Below

Scenario	Device							
	Audiovox 8910	Ericsson T68i	Sony Ericsson P910a	Motorola C333	Motorola V66	Motorola V300	Nokia 6610i	Nokia 6200
SMS/MMS Messaging	Miss	Below	Miss	Miss	Miss	Below	Below	Below
Internet Messaging	NA	Miss	Miss	NA	NA	Below	NA	NA
Web Applications	Miss	Miss	Miss	NA	NA	Miss	Miss	Miss
Text File Formats	NA	NA	Miss	NA	NA	Miss	Miss	Miss
Graphics File Formats	Miss	Miss	Miss	NA	NA	Miss	Miss	Miss
Compressed Archive File Formats	NA	NA	Miss	NA	NA	NA	Miss	Miss
Misnamed Files	NA	NA	Miss	NA	NA	NA	Miss	Miss
Peripheral Memory Cards	NA	NA	Miss	NA	NA	NA	NA	NA
Acquisition Consistency	NA	NA	NA	NA	NA	NA	NA	NA
Cleared Devices	NA	Meet	Meet	Meet	Meet	Meet	NA	NA
Power Loss	Above	Above	Meet	Above	Above	Above	Above	Above

SIM Card Acquisition

TULP2G version 1.2.0.2 gives examiners the ability to acquire SIM card data using a PC/SC-compatible reader. The acquisition steps followed to acquire data directly from the SIM are the same as acquiring data from a phone, except for selecting PC/SC Chip Card Communication instead of a serial or socket connection. The data fields acquired (e.g., Abbreviated Dialing Numbers, Last Numbers Dialed, Fixed Dialing Numbers, Messages) are dependent upon the SIM and service provider. The Report facilities operate in a similar fashion as for phone acquisitions, described above. Table 20 summarizes the results from applying the scenarios listed at the left of the table to the devices across the top. Additional information can be found in *NISTIR 7250 Cell Phone Forensic Tools: An Overview and Analysis*.

Table 20: SIM Card Results Matrix – External Reader

Scenario	SIM		
	5343	8778	1144
Basic Data	Meet	Meet	Meet
Location Data	Below	Below	Below
EMS Data	Meet	Meet	Below
Foreign Language Data	Meet	Meet	Meet

Synopsis of SecureView

SecureView version 1.5.0[15] available for both PC and Mac, has the ability to acquire data from various manufactures of GSM and non-GSM mobile devices. Device connectivity is established via either Susteen's unique cable interface, Bluetooth, or IR. Once connectivity has been established, the make and model of the device determines the richness of data (e.g., Phonebook, SMS, Image, Ringtone, Calendar) that can be recovered.

Supported Phone

SecureView supports over 300 cellular devices from the following manufacturers: Audiovox, Kyocera, LG, Motorola, Nokia, Samsung, Sanyo, Siemens and Sony Ericsson. The make, model, and type of phone determine the data elements (i.e., Phonebook, SMS, Images, Calendar) that SecureView supports, as illustrated below in Figure 28.

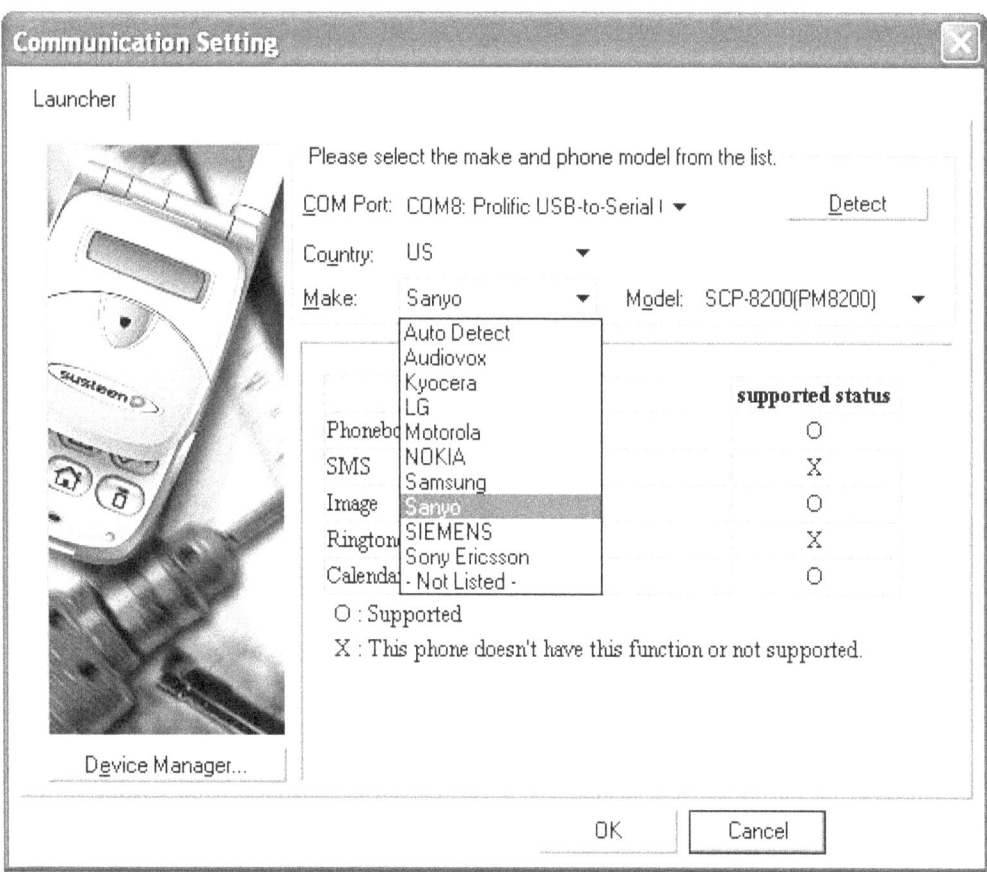

Figure 28: Supported Phone Selection

Acquisition Start

Acquisition begins by the examiner pressing the virtual keypad (i.e., tool icon) illustrated below in Figure 29. The tool icon key on the virtual keypad allows the examiner to decide whether the

---

[15] Additional information can be found at: **www.susteen.com**

device is auto-detected or manually selected via the make and model drop-down list. Once connectivity has been established with the device, the virtual keypad guides the examiner in selecting specific data elements to acquire. For instance, the music-note icon acquires ring-tones present on the device, if supported. Supported data elements that are acquired are presented in separate windows, specific to the data type (i.e., Phonebook, SMS, Image, Ringtone, Calendar), allowing examiners to view and examine the acquired data contents. Additional data (i.e., Phone Name, IMEI/ESN) can be determined by selecting the Properties tab within the Address Book as illustrated below in Figure 30.

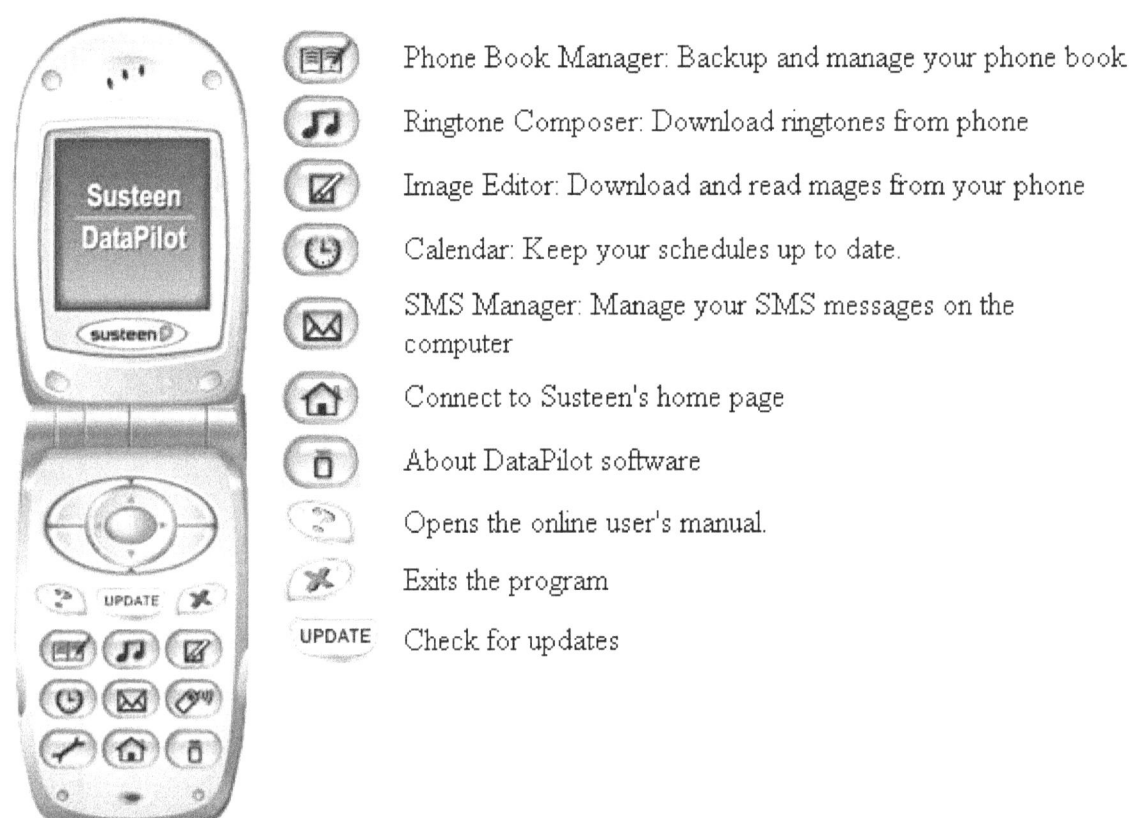

Figure 29: User Interface

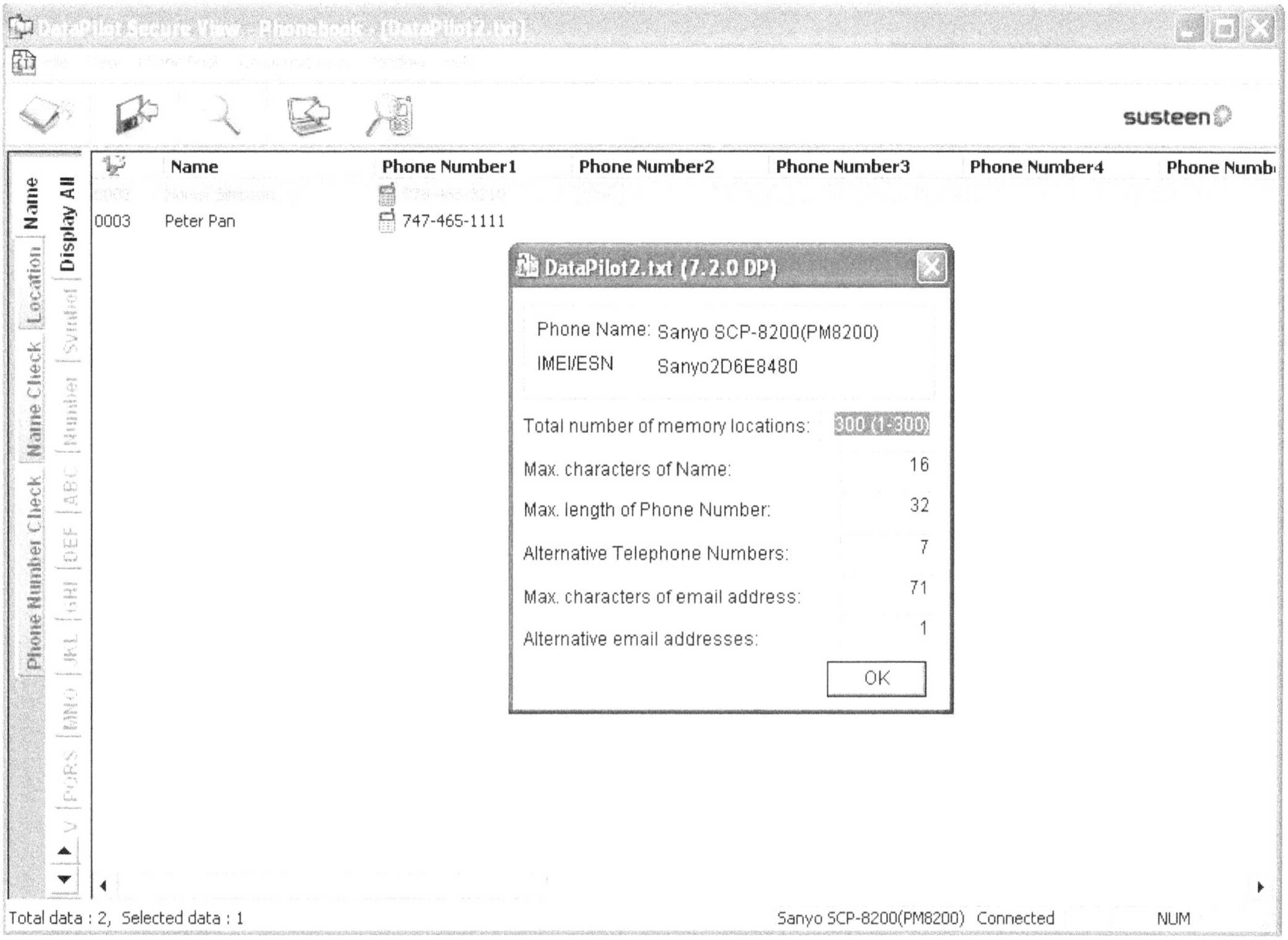

Figure 30: Phonebook Data

After data has been successfully acquired from the device, the data can be password protected through the Phonebook menu in the edit screen, illustrated below in Figure 31, to prevent unauthorized access to the case data.

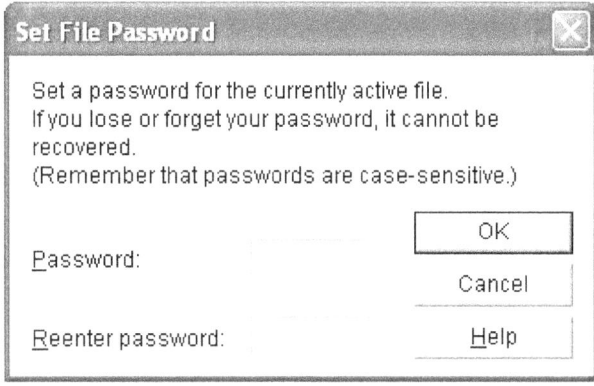

Figure 31: File Protection

Search Functionality

SecureView's search engine allows examiners to perform simplified string searches within specific acquired data windows (i.e., Phonebook, SMS, Image, Ringtone, Calendar). The search engine does not give examiners the ability to search through multiple cases or multiple folders within a case, or to issue complex expression patterns.

Graphics Library

The graphics library enables examiners to examine the collection of graphics files present on the device. Each image present can be viewed with a third-party application after it is saved to the hard disk. Figure 32 shows a screen shot of an image file and the corresponding folder.

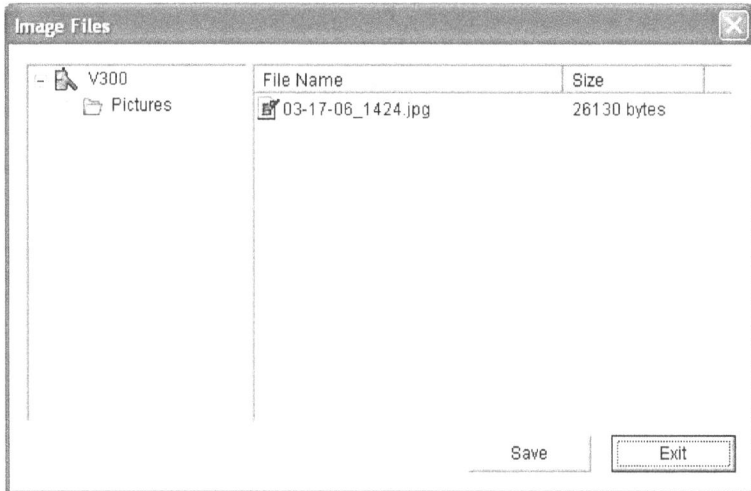

Figure 32: Image Files

Report Generation

SecureView does not support reporting facilities internally. Relevant data collected from the device can be exported and viewed with an appropriate third-party viewer. Third-party tools or editors can be used to create a finalized report of significant findings.

Scenario Results – Cell Phone

Table 21 summarizes the results from applying the scenarios listed at the left of the table to the devices across the top. More information can be found in Appendix D: SecureView Results.

Table 21: Results Matrix

Scenario	Device							
	Audiovox 8910	Ericsson T68i	LG 4015	Motorola C333	Motorola V66	Motorola V300	Nokia 6200	Sanyo 8200
Connectivity and Retrieval	Miss*	Meet	Meet	Meet	Meet	Meet		Meet
PIM Applications		Below	Below	Below	Below	Below		Below
Dialed/Received Phone Calls		Miss	Miss	Miss	Miss	Miss		Miss
SMS/MMS Messaging		Below	Miss	Below	Below	Below	Below	Miss
Internet Messaging		Miss	NA	NA	NA	Miss	NA	Miss
Web Applications		Miss	Miss	NA	NA	Miss	Miss	Miss
Text File Formats		NA	NA	NA	NA	Miss	Miss	NA
Graphics File Formats		Miss	NA	NA	NA	Meet	Below	Below
Compressed Archive File Formats		NA	NA	NA	NA	NA	Miss	NA
Misnamed Files	NA	NA	NA	NA	NA	NA	Miss	NA
Peripheral Memory Cards		NA	NA	NA	NA	NA	NA	NA
Acquisition Consistency		NA	NA	NA	NA	NA	NA	NA
Cleared Devices		Above	Above	Meet	Meet	Meet	NA	Above
Power Loss		Above	Above	Above	Above	Above	Above	Above

Scenario Results – SIM Card Acquis...

SecureView version 1.5.0 gives examiners the ability to acquire SIM card data using a PC/SC-compatible reader. The acquisition steps followed to acquire data directly from the SIM are the same as acquiring data from a phone, except for selecting PC/SC Chip Card Communication. The data fields acquired (e.g., Abbreviated Dialing Numbers, Last Numbers Dialed, Fixed Dialing Numbers, Messages, etc.) are dependent upon the SIM and service provider. The Report facility operates in a similar fashion to phone acquisitions, described above.

Table 22 summarizes the results from applying the scenarios listed at the left of the table to the SIMs across the top. More information can be found in Appendix I: SecureView – External SIM Results.

---

* Acquisition is supported only for non pay-as-you go carriers

Table 22: SIM Card Results Matrix - External Reader

Scenario	SIM		
	5343	8778	1144
Basic Data	Below	Below	Below
Location Data	Miss	Miss	Miss
EMS Data	Miss	Miss	Miss
Foreign Language Data	Below	Below	Below

Synopsis of PhoneBase2

PhoneBase2 version 1.2.0.15[16] has the ability to acquire data from various manufactures of GSM and non-GSM mobile devices. Device connectivity is established via third-party cable solutions, Bluetooth, or infrared. The cellular device acquisition engine used in PhoneBase2 is apparently licensed from the manufacturers of MobileEdit!. Comparable results could be expected when the tools are applied to the same device.

PhoneBase2's user interface provides examiners with quick and easy access to all of the major system functions. Data retrieved from cellular devices or associated media are stored in a common database format. Users can be assigned "Authority Profiles" which grants or deny the use of specific functions such as, modifying existing records, deleting records, or initiating data reads. Examiners have the ability to analyze, examine, search, store, bookmark, and create customized reports for acquired data. The "Main screen", illustrated below in Figure 33, contains four tabs providing examiners quick access to recently created case files (Today), all case files (Index), groupings of records based on date created, user, category, and status (Folder), and search capabilities within a database (Search).

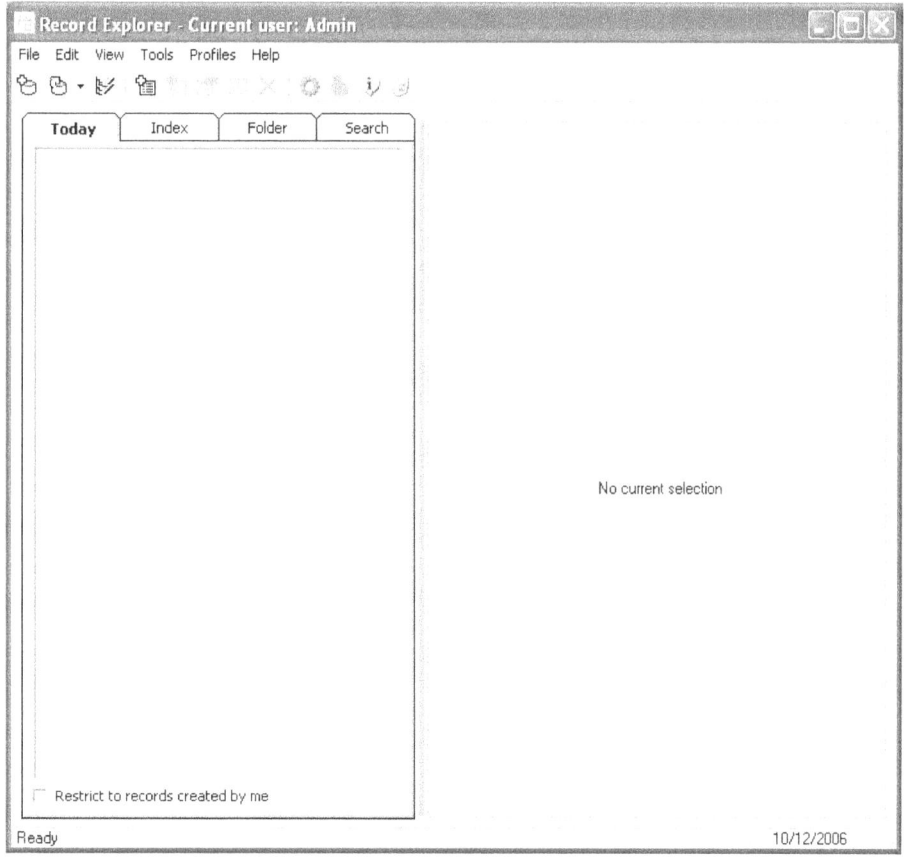

Figure 33: PhoneBase2 - User Interface

---

[16] Additional information can be found at: **www.phonebase.info**

Supported Phones

PhoneBase2 supports over 200 cellular devices from the following manufacturers: Alcatel, Ericsson, LG, Motorola, Nokia, Panasonic, Philips, Samsung, Siemens and Sony Ericsson. The make, model, and type of phone determine the data elements (i.e., Phonebook, SMS, Images, Calendar) that SecureView supports.

Acquisition Stage

The acquisition process begins by successfully logging into the PhoneBase2 application via either Admin or an assigned username, creating a case (e.g., File -> New) and adding a unique record to a database provided with the PhoneBase2 software, as illustrated below in Figure 34 and Figure 35.

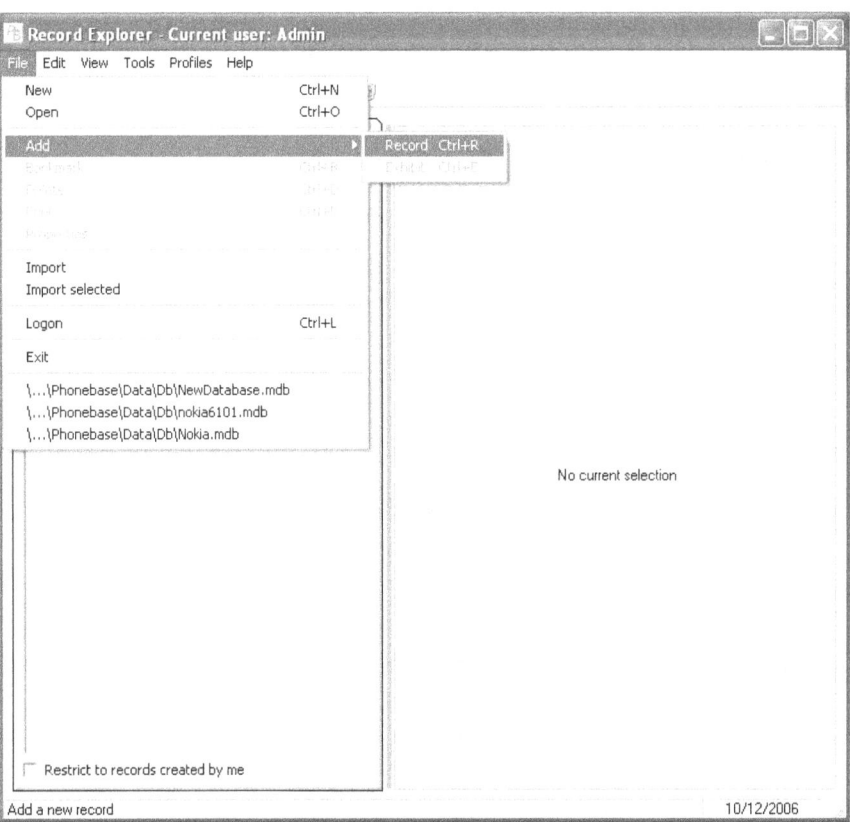

Figure 34: PhoneBase2 - Adding Records

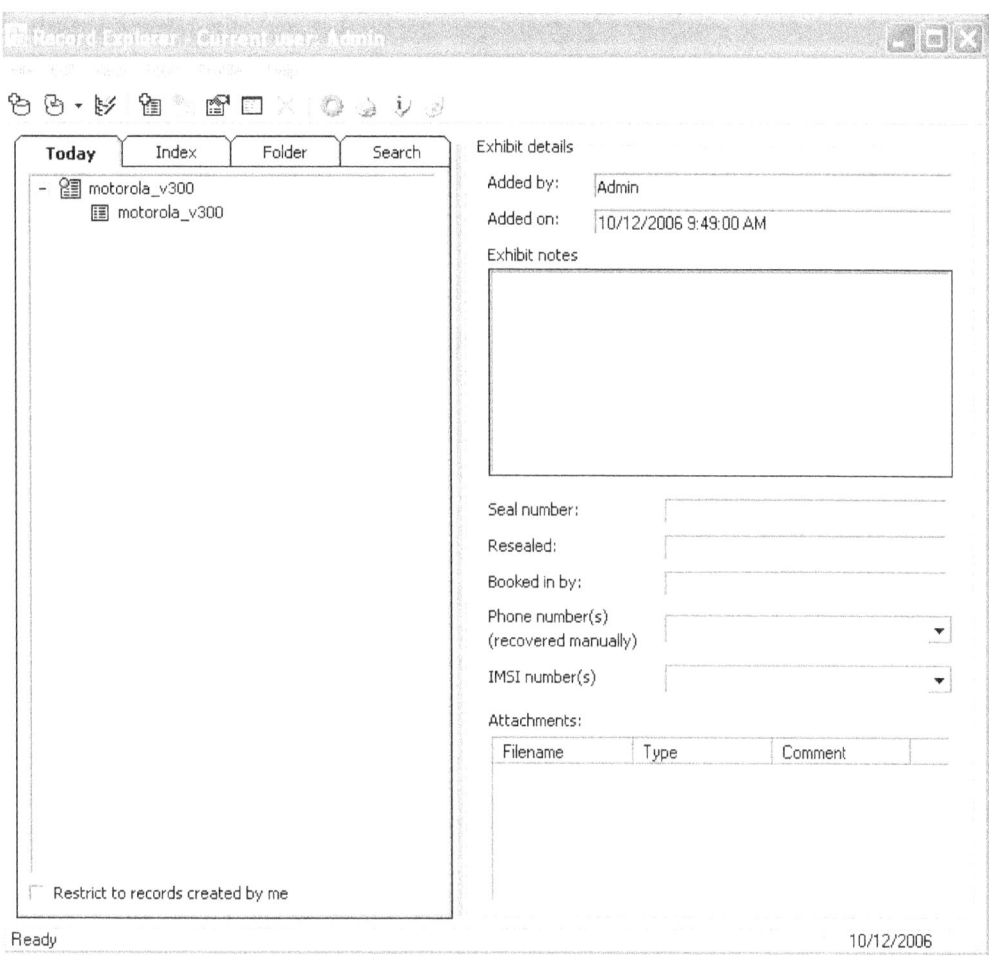

Figure 35: PhoneBase2 - Acquisition

After case notes and the appropriate fields containing make and model of the device are entered the READ button will appear, which initiates connectivity to the device and the reading of data as illustrated below in Figure 36.

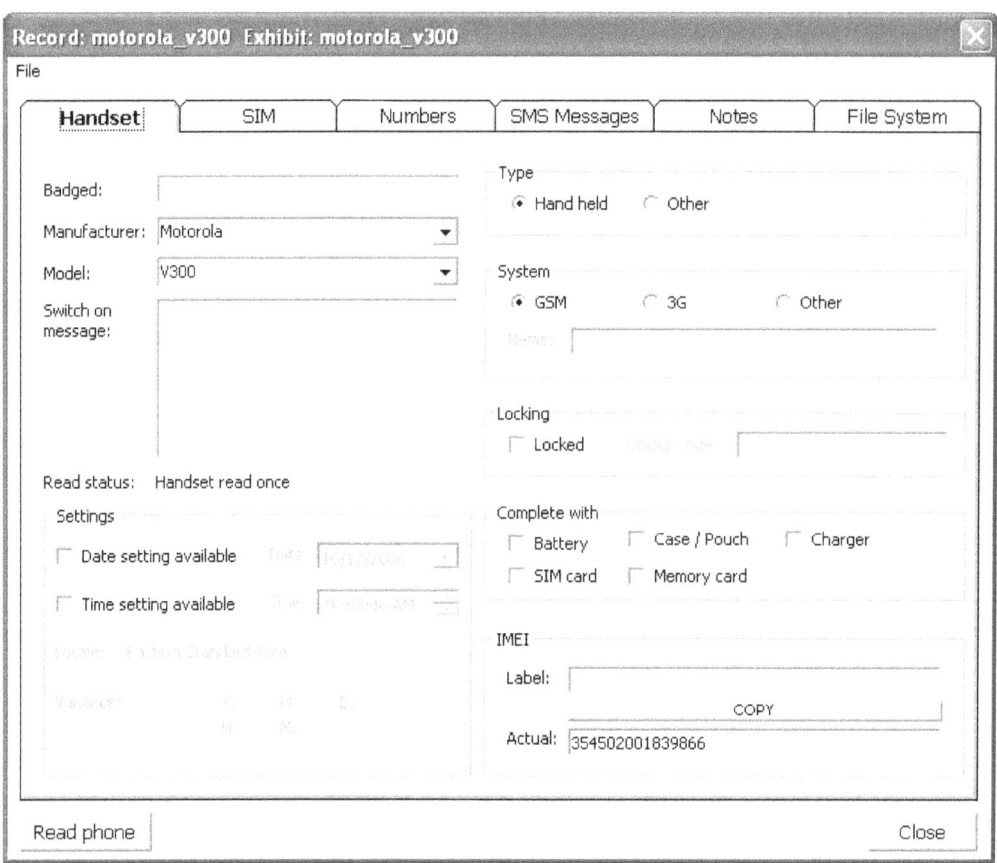

Figure 36: PhoneBase2 - Read phone

After the device contents have been successfully acquired, the PhoneBase2 user-interface guides examiners via data descriptors located on the content tabs as illustrated in Figure 37 and Figure 38 below.

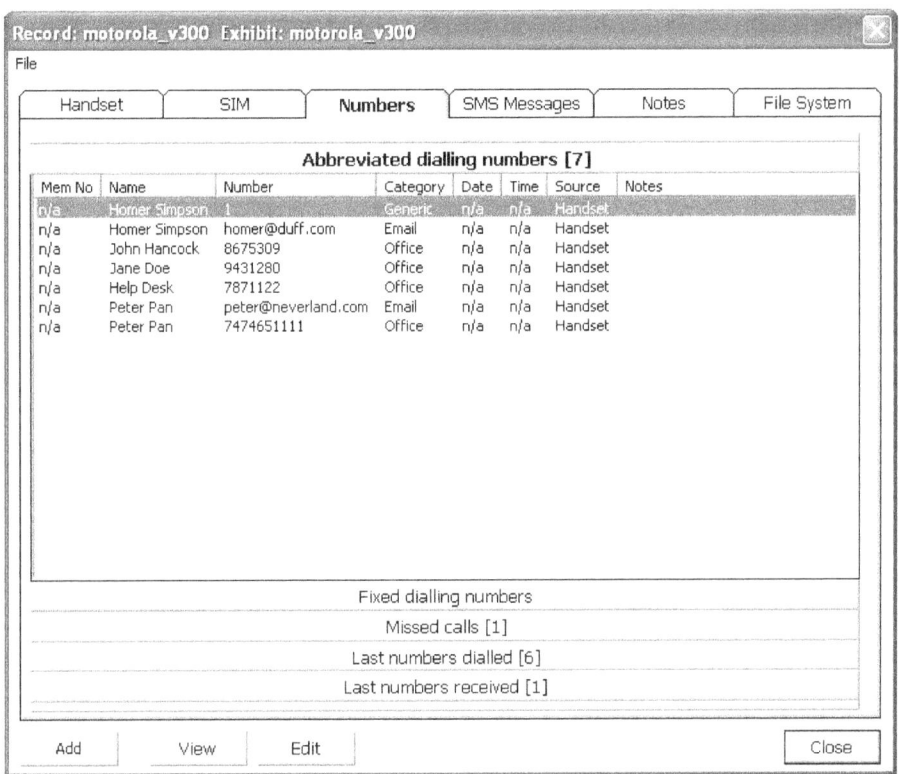

Figure 37: PhoneBase2 - Address Book

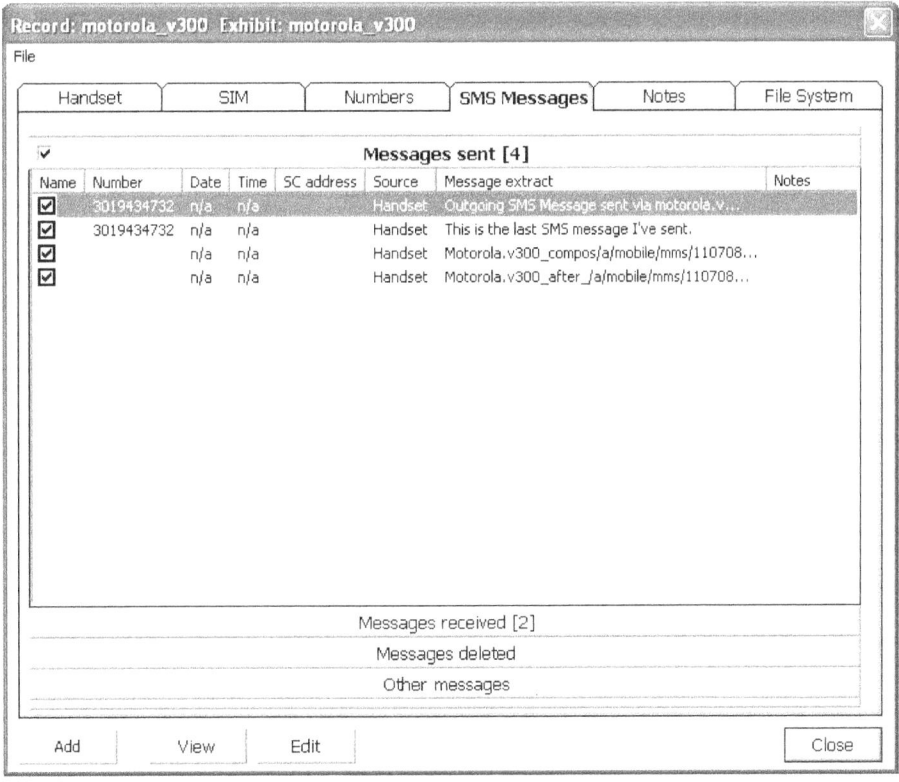

Figure 38: PhoneBase2 - Text Messages

Search Functional

PhoneBase2 provides examiners with the option to execute a simple query on 15 fields (i.e., Record ID, Exhibit ID, Case Notes, SIM card notes, IMSI number, Serial number (ME), Serial number (ICC), Phone number, Numbers, Message text, Contact name, Network provider, Attachment, Exhibit Notes, Seal number). PhoneBase2 pre-selects the first three before-mentioned fields by default. When a database search is processed, all cases that contain the queried search string will be displayed. The maximum limit for search returns is 200.

Graphics Library

Beneath the filesystem tab rests the Graphics, Video, Audio, and Other sub-tabs that contain any associated files acquired from the device of the type designated by the sub-tab. Illustrated in Figure 39 is the Images sub-tab that enables examination of the collection of graphics files present on the device. Each image present can be viewed internally, by clicking the Edit button, or can be exported and inspected with a third party tool.

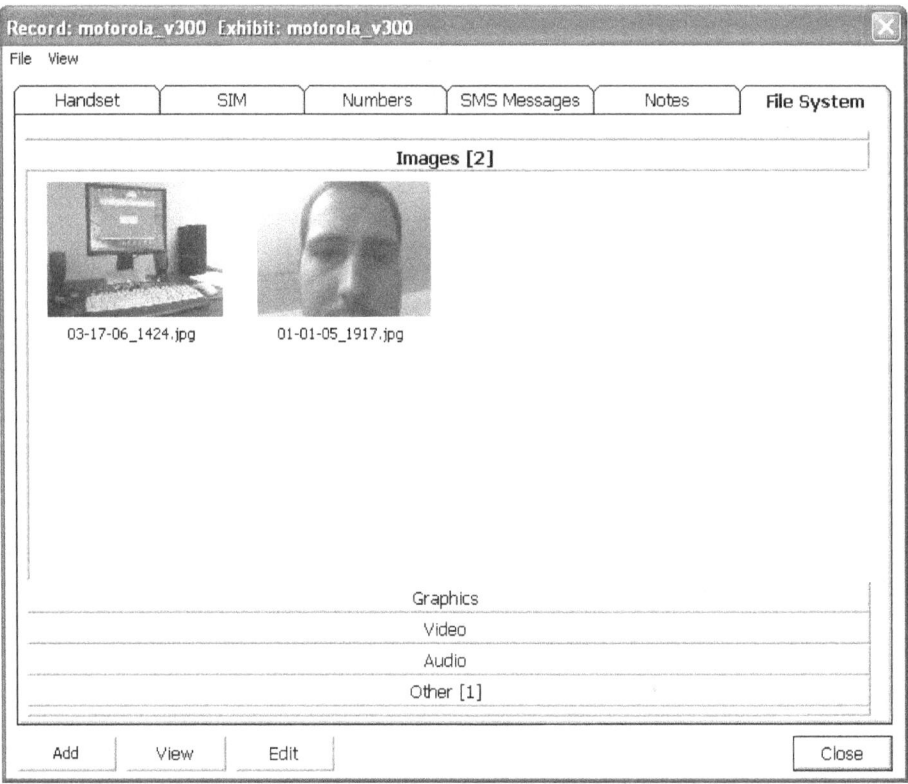

Figure 39: PhoneBase2 - Graphics Library

Report Generation

PhoneBase2 provides a report generation facility that allows the creation of a hard-copy or soft-copy report (using third party software) that summarizes relevant data found on the device. Illustrated below in Figure 40 are data items that can be selected for reporting purposes. PhoneBase2 does not provide examiners with the option to include additional case data such as graphic files in the finalized report.

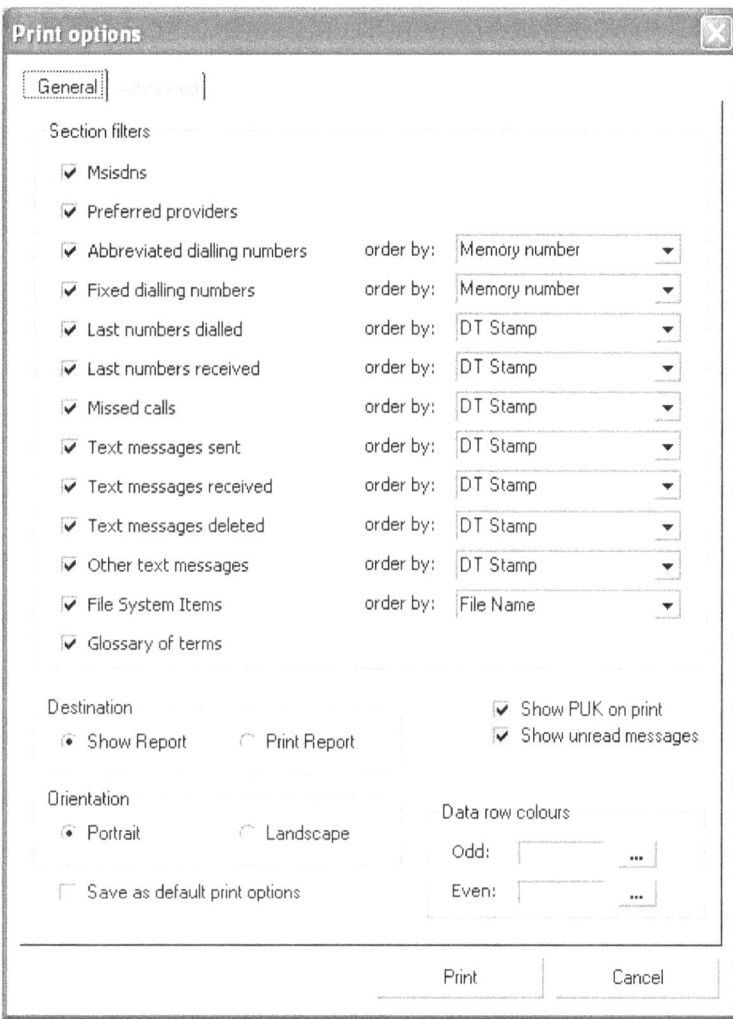

Figure 40: PhoneBase2 - Print options

Scenario Results – Cell Phones

Table 23 summarizes the results of applying the scenarios listed at the left of the table to the devices listed across the top. More information can be found in Appendix E: PhoneBase2 Results.

Table 23: Results Matrix

Scenario	Device			
	Ericsson T68i	Motorola v66	Motorola v300	Nokia 6610i
Connectivity and Retrieval	Meet	Meet	Meet	Meet
PIM Applications	Below	Below	Below	Below
Dialed/Received Phone Calls	Below	Below	Below	Below

Scenario	Device			
	Ericsson T68i	Motorola v66	Motorola v300	Nokia 6610i
SMS/MMS Messaging	Below	Below	Below	Below
Internet Messaging	Miss	NA	Miss	NA
Web Applications	Miss	NA	Miss	Miss
Text File Formats	NA	NA	Miss	Meet
Graphics File Formats	Miss	NA	Meet	Below
Compressed Archive File Formats	NA	NA	NA	Meet
Misnamed Files	NA	NA	NA	Meet
Peripheral Memory Cards	NA	NA	NA	NA
Acquisition Consistency	NA	NA	NA	NA
Cleared Devices	Meet	Meet	Meet	NA
Power Loss	Above	Above	Above	Above

Scenario Results - SIM Card Acquis

PhoneBase2 version 1.2.0.15 gives examiners the ability to acquire SIM card data using a PC/SC-compatible reader. The acquisition steps followed to acquire data directly from the SIM are the same as acquiring data from a phone, except for selecting PC/SC Chip Card Communication. The data fields acquired (e.g., Abbreviated Dialing Numbers, Last Numbers Dialed, Fixed Dialing Numbers, Messages) are dependent upon the SIM and service provider. The Report facility operates in a similar fashion as for phone acquisitions, described above.

Table 24 summarizes the results from applying the scenarios listed at the left of the table to the SIMs across the top. More information can be found in Appendix J: PhoneBase2 – External SIM Results.

Table 24: SIM Card Results Matrix - External Reader

Scenario	SIM		
	5343	8778	1144
Basic Data	Below	Below	Below
Location Data	Below	Below	Below
EMS Data	Below	Below	Below
Foreign Language Data	Below	Below	Below

Synopsis of CellDEK

CellDEK version 1.3.1.0 has the ability to acquire data from various manufactures of GSM and non-GSM mobile devices. Device connectivity is established via either CellDEK's unique cable interface, Bluetooth, or IR. The make and model of the device determines the data (e.g., Phonebook, SMS, Image, Ringtone, Calendar) that CellDEK recovers. All data is individually hashed using the MD5 algorithm to ensure that the integrity of the data can be verified. Additionally, CellDEK generates detailed log files of the acquisition process, which captures every button press, all commands sent to the target device and all data received from the target device.

Supported Phones

CellDEK version 1.3.1.0 supports approximately 220 cellular devices. The following manufacturers are supported: Acer, BenQ, BenQ Siemens, Blackberry, HP, iMATE, Kyocera, LG, Motorola, Nokia, O2, Orange, Samsung, Siemens, and Sony Ericsson.

Acquisition Start

The initial or startup screen for the CellDEK application is displayed below in Figure 41. This menu provides the examiner with the ability to do various activities (e.g., acquisition, viewing data, backing up files, upgrading software) and serves as the main navigational page. The acquisition process begins by selecting "read new device".

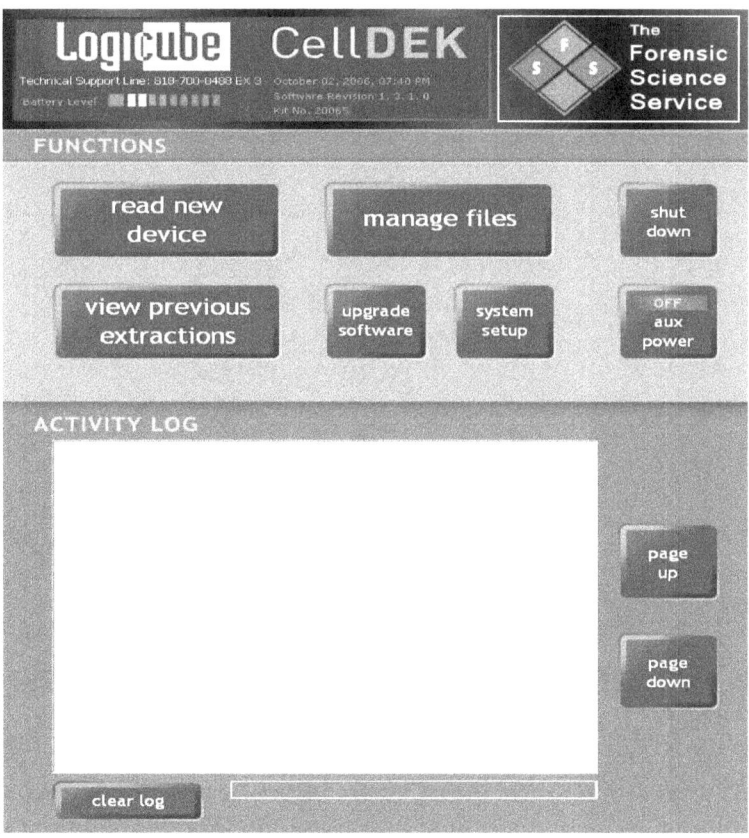

Figure 41: User Interface

The acquisition phase continues by prompting the examiner to select which type of device or media (i.e., Cell Phone, PDA, SIM Card, Flash Memory) to acquire, as illustrated below in Figure 42. After the device type has been chosen (for this illustration – "Cell Phone" was selected), the examiner is asked to choose the device manufacturer as displayed in Figure 43.

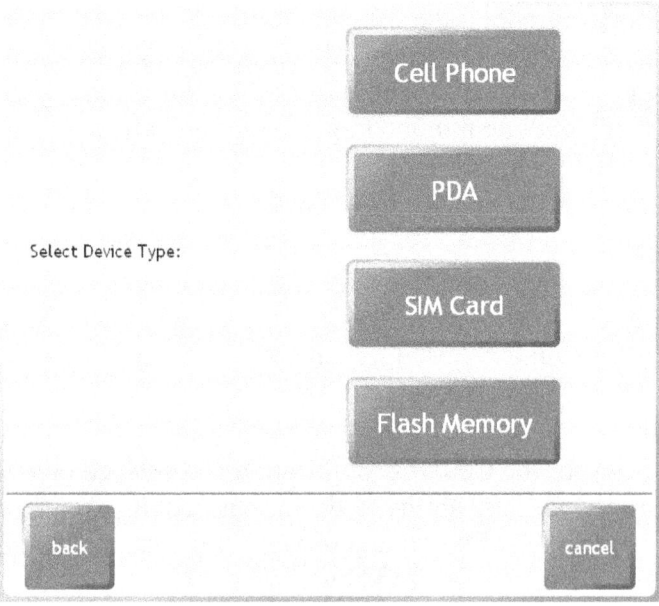

Figure 42: CellDEK - Device Type

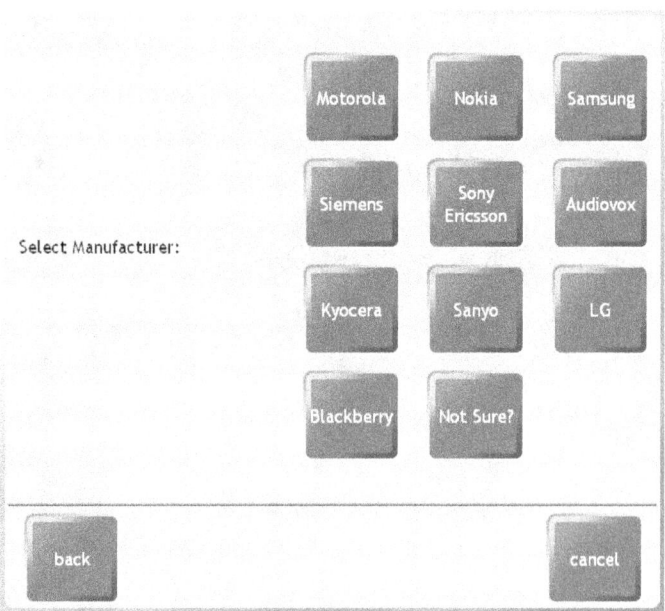

Figure 43: CellDEK - Device Manufacturer

In order for acquisition to continue and the proper cable to be identified, the model of the device must be entered as illustrated below in Figure 44. If the model is unknown or cannot be determined, CellDEK provides a graphical representation of all supported devices for aiding identification. Additionally, if this proves to be unsuccessful, the dimensions (i.e., length, width, depth) of the device can be entered and CellDEK will display potential devices that fall within those measurements are displayed.

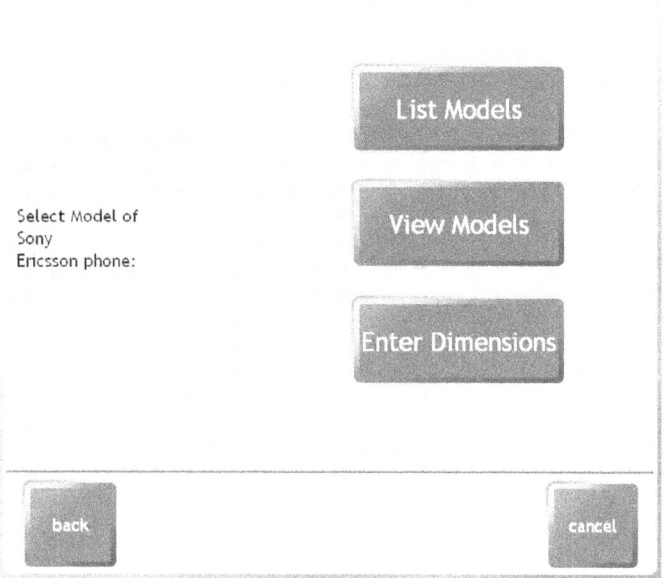

Figure 44: CellDEK - Model Selection

Once the device has been identified, the application illuminates the recommended connectivity procedure (i.e., cable, IR, or Bluetooth) and the data types able to be recovered, as illustrated below in Figure 45. If a cable connection is allowable, the appropriate cable connector lights up, alerting the examiner to plug the lit connector into the device bed. Once the cable has been seated properly into the device bed, the examiner connects the cable to the phone interface. If IR or Bluetooth connections are utilized, specific instructions are displayed informing the examiner on wireless acquisition instructions.

Figure 45: CellDEK - Connectivity

When the acquisition procedure completes, the "DATA VIEW" interface illustrated below in Figure 46 is presented. This interface provides the means to select and view the acquired data elements such as Device Info, Contacts, Call Log, Text Messages, Organizer, and Extended Data. The latter contains recorded images, audio and video. The illustration below gives an example screenshot of the Contacts selection and associated data present on the acquired device.

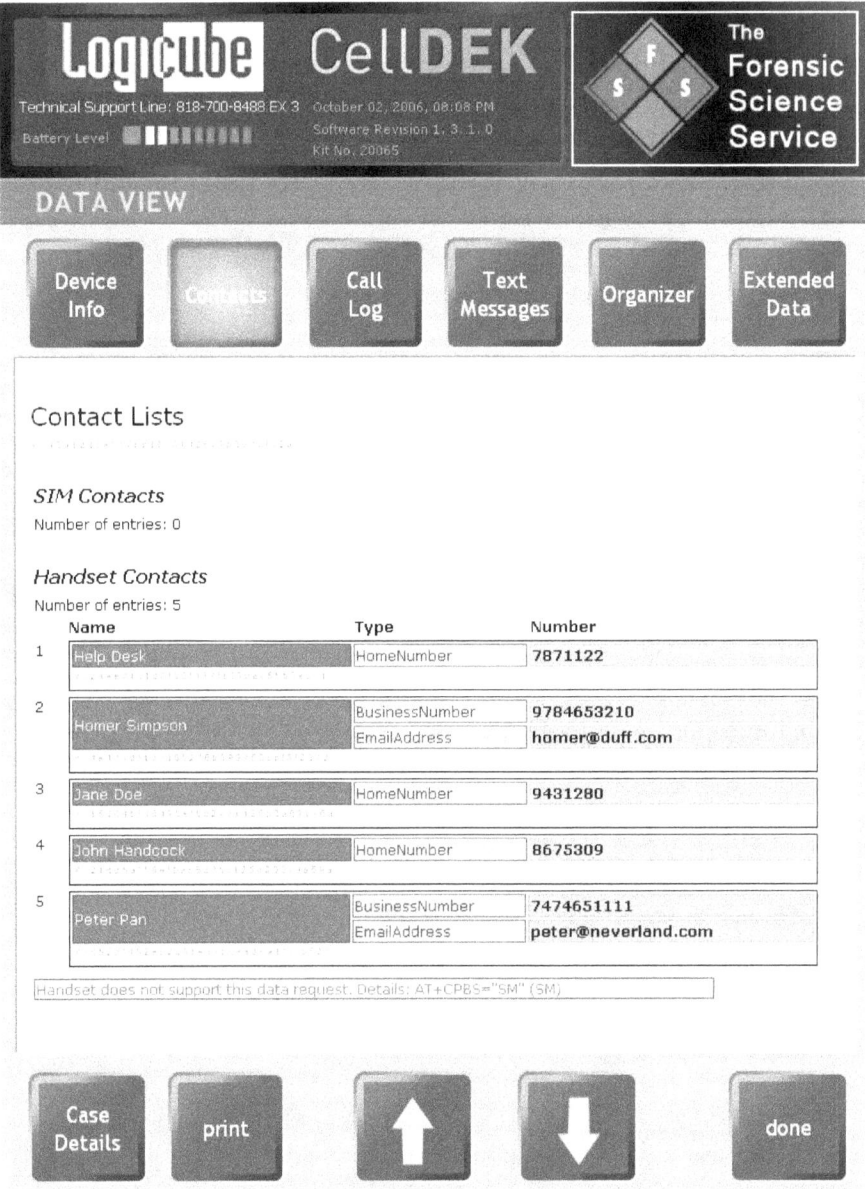

Figure 46: CellDEK - Contacts

Search Functionality

The CellDEK unit does not have any search functionality built into the application. Generated reports can be searched for specific strings using a third party tool.

Graphics Library

CellDEK classifies graphics files as well as video and audio files, as "Extended Data". Once the standard data (i.e., Contacts, Call Log, Text Messages, Organizer) has successfully been extracted, the option to download extended data found on the device or media is given. Graphic file data is displayed as illustrated below in Figure 47 (thumbnails can be expanded by tapping the image with the stylus).

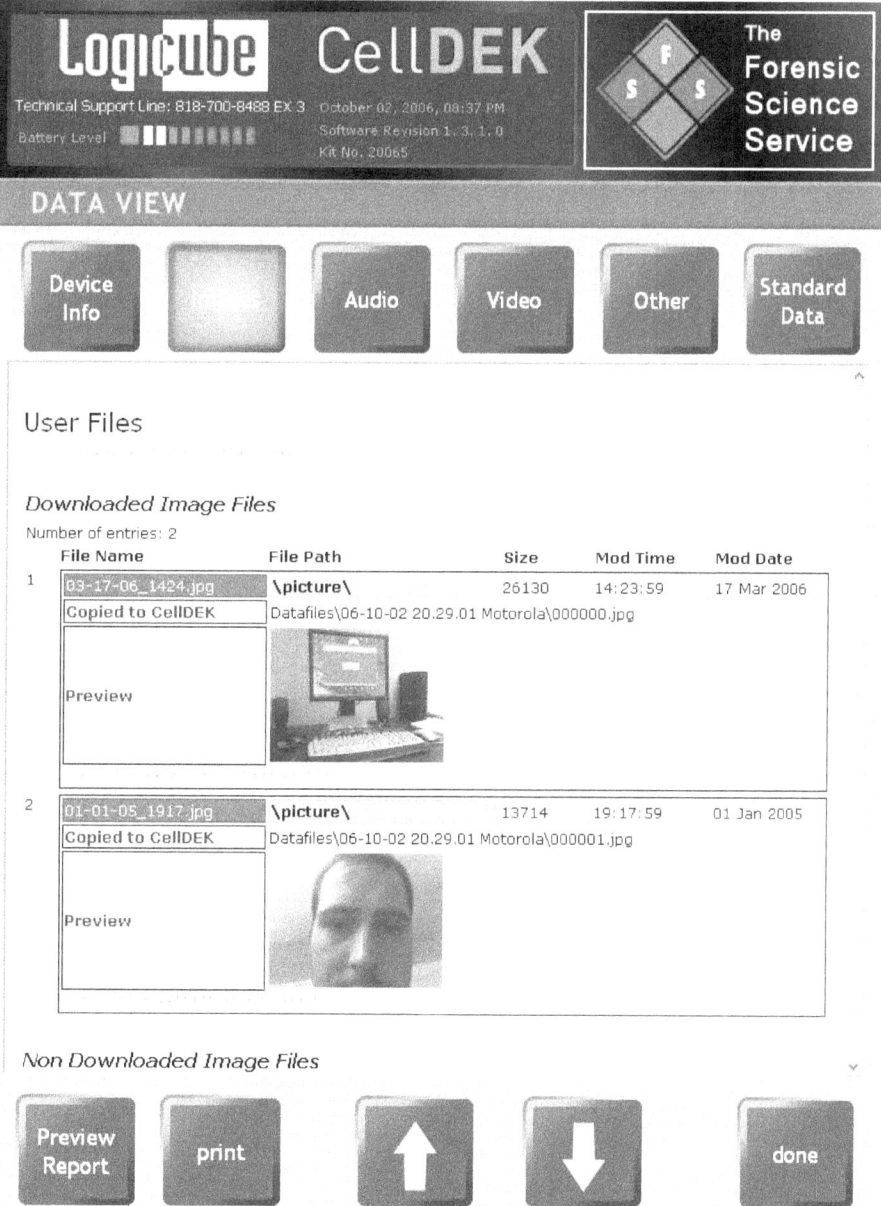

Figure 47: CellDEK - Graphics Library

Report Generation

CellDEK automatically generates report files containing all of the data found in an HTML format, via the user interface. Reports can be customized with company logos and case details when entered by the examiner prior to device acquisition as shown in Figure 48.

Figure 48: CellDEK - Case Details

Scenario Results – Cell Phone

Table 25 summarizes the results of applying the scenarios listed at the left of the table to the devices listed across the top. More information can be found in Appendix F: CellDEK Results.

Table 25: Results Matrix

Scenario	Device					
	Ericsson T68i	Motorola MPX220	Motorola V66	Motorola V300	Nokia 6200	Nokia 6610i
Connectivity and Retrieval	Meet	Meet	Meet	Meet	Meet	
PIM Applications	Below	Below	Below	Below	Below	Below
Dialed/Received Phone Calls	Below	Below	Below	Below	Below	Below
SMS/MMS Messaging	Below	Below	Below	Below	Below	Below

78

Scenario	Device					
	Ericsson T68i	Motorola MPX220	Motorola V66	Motorola V300	Nokia 6200	Nokia 6610i
Internet Messaging	Miss	Below	NA	Below	NA	NA
Web Applications	Miss	Miss	NA	Miss	Miss	Miss
Text File Formats	NA	Miss	NA	Miss	Miss	Miss
Graphics File Formats	Miss	Below	NA	Meet	Below	Below
Compressed Archive File Formats	NA	Miss	NA	NA	Miss	Miss
Misnamed Files	NA	Miss	NA	NA	Miss	Miss
Peripheral Memory Cards	NA	Below		NA	NA	NA NA
Acquisition Consistency	Meet	Meet	Meet	Meet	Meet	Meet
Cleared Devices	NA	Meet	Meet	Meet	NA	NA
Power Loss	Above	Above	Above	Above	Above	Above

Scenario Results - SIM Card Acquis

CellDEK version 1.3.1.0 provides the ability to acquire SIM card data using the PC/SC-compatible reader built into the right hand side of the carrying case. A green LED flashes, when the SIM card is seated correctly. The acquisition process begins by tapping the "SIM Card" button on the Select Device Type Screen, which prompts the examiner to insert the SIM into the reader as illustrated below in Figure 49.

Figure 49: CellDEK - SIM Acquisition

During the acquisition process, an activity log is displayed, as illustrated below in Figure 50, providing details of what data items are currently being acquired from the SIM. Prior to the reading of the SIM, the examiner is prompted with the "Case Details" screen allowing additional notes about the data recovered from the SIM to be entered via the virtual keyboard. The Case Details are included in the final report as described above.

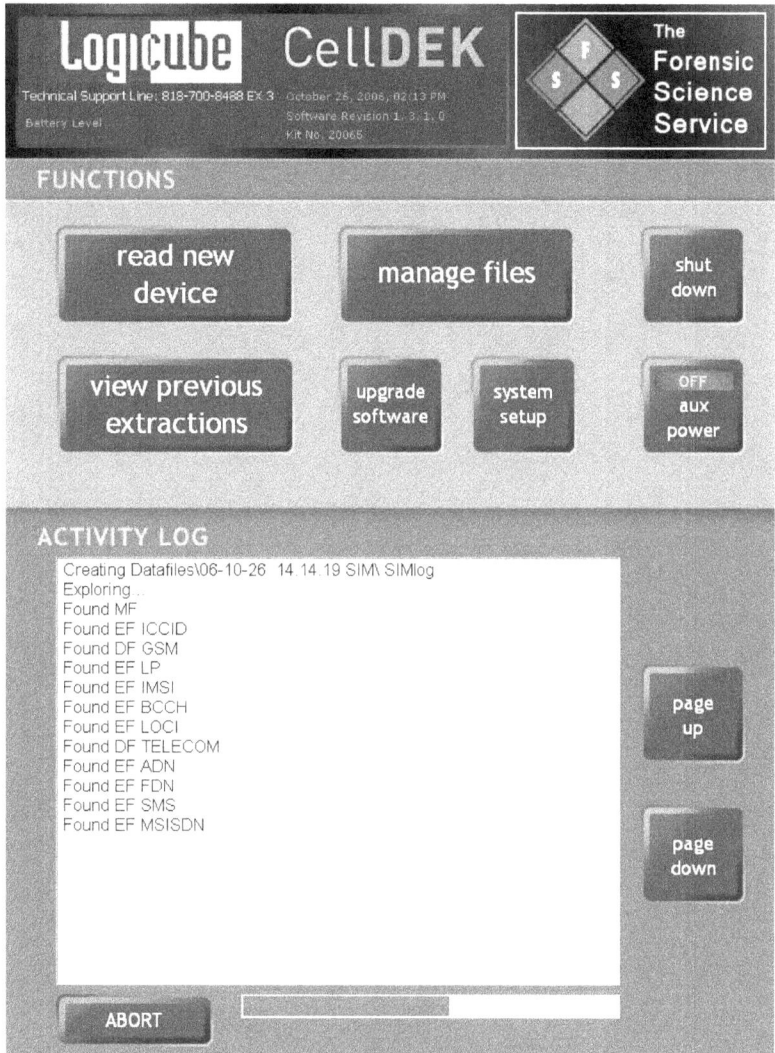

Figure 50: CellDEK - Activity Log

The following data fields are reported: IMSI, ICCID, MSISDN, ADN, LND, SMS, and LOCI information. Table 26 summarizes the results of applying the scenarios listed at the left of the table to the SIMs across the top. More information can be found in Appendix K: CellDEK – External SIM Results.

Table 26: SIM Card Results Matrix - External Reader

Scenario	SIM		
	5343	8778	1144
Basic Data	Below	Below	Below
Location Data	Miss	Miss	Miss
EMS Data	Below	Below	Below
Foreign Language Data	Below	Below	Below

80

Synopsis of SIMIS2

SIMIS2 version 2.2.11 from Crown Hill[17] is able to acquire information from SIM cards via the PC/SC-compatible card reader that comes with the software. SIMIS2 provides examiners with a user interface containing a set of tabs providing examiners with the ability to create report notes, import archived reports, search acquired data and perform PIN administration. SIMIS2 allows the following data types to be acquired from SIM cards: Abbreviated Dial Numbers (ADN), Fixed Dial Numbers (FDN), International Mobile Subscriber Identity (IMSI), Last Numbers Dialed (LND), Mobile Subscriber Integrated Services Digital Network Number (MSISDN), Short Message Server Parameters (SMSP), SMS Short Messages, Deleted Messages, Public Land Mobile Networks (PLMNS), Forbidden Public Land Mobile Networks (FPLMNS), Location Information, Broadcast Control Channel (BCCH), Cell Broadcast Message Identifier for Data Download (CBMID), Voicemail Number, Integrated Circuit Card Identification (ICCID), Phase ID, Service Provider, Administration Data, Service Dialing Number (SDN) and Capability Configuration Parameters. The tool can also perform a full dump of the card contents for analysis.

The acquisition begins by collecting case information (i.e., unique file reference, operator name, case number, IMEI, ICCID and Service Provider information imprinted on the SIM) from the examiner for inclusion in the final report. After a successful acquisition, SIMIS2 compares the data manually entered (e.g., the ICCID) with the data acquired from the SIM for consistency. If a discrepancy is found, the examiner is notified. SIMIS2 allows examiners to view additional data that is not displayed within the user interface. An ASCII dump file (.dmp) can be created in the SIMIS2 output directory by selecting "SIM dump" from the File menu. For example, the Forbidden Networks (FPLMNs) elementary file can be obtained this way.

Generated reports can be customized with specifics related to the examination before acquisition occurs, which are included in the final version. After selecting the "Read SIM" option and acquiring all data, the report is generated is displayed to the user. The generated report is an HTML-based report that can be viewed with a standard HTML editor. The SIMIS2 Search engine provides examiners the ability to perform searches on phone numbers contained within a case or across the entire database of all cases created. Currently, alpha string searches are not supported. SIMIS2 currently does not display any graphics found in certain Extended SMS messages.

Table 27 summarizes the results from applying the scenarios listed at the left of the table to the devices across the top. Additional information can be found in *NISTIR 7250 Cell Phone Forensic Tools: An Overview and Analysis*.

Table 27: SIM Card Results Matrix – External Reader

Scenario	SIM		
	5343	8778	1144
Basic Data	Meet	Meet	Meet

---

[17] For additional information on Crownhill USA products see: **www.crownhillmobile.com**

Scenario	SIM		
	5343	8778	1144
Location Data	Meet	Below	Below
EMS Data	Below	Below	Below
Foreign Language Data*	Meet	Meet	Meet

---

* Previous versions of SIMIS2 did not correctly display Unicode characters. Version 2.2.11 has corrected this issue.

Synopsis of ForensicSIM

The ForensicSIM toolkit from Radio Tactics is able to acquire information from SIM cards via a PC/SC-compatible card reader that comes with the software. Before the data contents of the SIM can be analyzed with the Forensic Analysis software, the examiner must use the ForensicSIM acquisition terminal, a standalone device, to create a reference copy of the target SIM on a separate storage card. To copy the SIM, the examiner first logs on to the acquisition terminal with a username and PIN number. The acquisition terminal then walks the examiner through the process of entering the target SIM and creating duplicate reference copies on the provided blank SIM cards. A Master SIM is intended for storage in case of an evidence dispute, while a Prosecution SIM copy that serves as a working duplicate for evidence recovery and analysis, and a Defense SIM copy that serves as a working duplicate issued to the defense. In addition to the aforementioned SIM copy cards, an option exists to create an Access Card. The Access Card is a copy of the target SIM which can be inserted into an associated GSM handset for examination without the risk of connecting to the cell network.

Once the target SIM has been successfully duplicated, the ForensicSIM software and SIM reader can be used to create a report of the data contained on a reference SIM. The ForensicSIM toolkit allows the following data types to be recovered: Subscriber/User related files, Phone Number related files, SMS related files, Network related files and General SIM information. Each individual field contains additional meta-data about each type.

The examiner has the option of creating a standard report or an advanced report. A standard report displays only card identification information, phonebook, and SMS text messages. The advanced report displays additional information recovered from the SIM card. After the data on the SIM has been successfully acquired, the examiner is then asked to enter case-specific information (i.e., Operator, Operator Name, Date/Time, Reference No., Case Reference No., Case Officer, Exhibit Reference, Exhibit Seal No., Exhibit Reseal No., Phone Make, Phone Type, IMEI and PIN/PUK codes, if known). After the report is generated, examiners can manually search the data it contains or use an automated search engine appropriate for the exported file type of the report. Table 28 summarizes the results from applying the scenarios listed at the left of the table to the devices across the top. Additional information can be found in *NISTIR 7250 Cell Phone Forensic Tools: An Overview and Analysis*.

Table 28: SIM Card Results Matrix - External Reader

Scenario	SIM		
	5343	8778	1144
Basic Data	Meet	Miss	Below
Location Data	Meet	Miss	Below
EMS Data	Below	Miss	Below
Foreign Language Data	Below	Miss	Below

Synopsis of Forensic Card Reader

Forensic Card Reader (FCR)[18] version 1.8.0.94 is able to acquire information from SIM cards via the PC/SC Chipy reader. The FCR PC/SC USB card reader and FCR software give examiners the ability to capture data such as the ICC ID, IMSI, incoming/outgoing calls, abbreviated call numbers, SMS messages and location data. Data elements and acquisition progress are displayed, allowing the acquisition process to be monitored. FCR does not have a built-in search engine or search facilities. Once the XML-formatted report is generated, examiners can manually search the data it contains or use an appropriate search tool. FCR does not display graphics files of any type. The FCR application only allows examiners to export textual data in the XML format. Once the finalized report is generated, the file can be viewed with an appropriate Web browser or XML editor.

Table 29 summarizes the results from applying the scenarios listed at the left of the table to the devices across the top. Additional information can be found in *NISTIR 7250 Cell Phone Forensic Tools: An Overview and Analysis*.

Table 29: SIM Card Results Matrix - External Reader

Scenario	SIM		
	5343	8778	1144
Basic Data	Below	Below	Below
Location Data	Below	Below	Below
EMS Data	Below	Below	Below
Foreign Language Data	Below	Below	Below

---

[18] Additional information on FCR can be found at: **http://www.becker-partner.de/forensic/intro_e.htm**

Synopsis of SIMCon

SIMCon version 1.2 can acquire information from SIM cards via a PC/SC-compatible reader. The SIMCon software gives examiners the ability to capture and examine data such as the Card Identity (ICCID), Stored Dialing Numbers (ADN), Fixed Dialing Numbers (FDN), Subscriber Number (MSISDN), Last Numbers Dialed (LND), SMS Messages, Subscriber Identity (IMSI), Ciphering Key (Kc) and Location Information (LOCI). After a successful acquisition, the entire SIM contents can be saved and stored in the SIMCon .sim proprietary format for later processing. SIMCon allows searching for non-standard "hidden files" by checking the "Search for hidden files" checkbox. SIMCon uses an internal hashing facility to ensure the integrity of cases and detect whether tampering occurred during storage. SIMCon uses the SHA1 algorithm to compute a hash for each file as it is read from the card. Selecting "Verify Hash" in the "File" menu causes SIMCon to recompute all hashes and check that the original file is consistent with the reopened case.

SIMCon does not have a built-in search engine or search facilities. However, data can be viewed manually by browsing through the tree structure and selecting individual data items. A textual and hexadecimal representation of a selected data item is provided in the lower pane. Also, once the formatted report is generated, examiners can manually search it manually or use an appropriate automated search tool. SIMCon only supports small (16x16) and large (32x32) pixel images embedded in certain Extended SMS messages. These types of images are reported alongside the textual content of the message. Report generation begins by prompting the examiner with case-specific details such as investigator name, date/time, case id, evidence number, and notes specific to the investigation. The report includes the case data entered and any data items selected by the examiner for inclusion.

Table 30 summarizes the results from applying the scenarios listed at the left of the table to the devices across the top. Additional information can be found in *NISTIR 7250 Cell Phone Forensic Tools: An Overview and Analysis*.

Table 30: SIM Card Results Matrix - External Reader

Scenario	SIM		
	5343	8778	1144
Basic Data	Meet	Below	Below
Location Data	Meet	Meet	Meet
EMS Data	Meet	Meet	Meet
Foreign Language Data	Meet	Meet	Meet

Synopsis of USIMdetective

Quantaq's USIMdetective version 1.3.1[19] can acquire information from SIM cards via a PC/SC-compatible reader. The USIMdetective software provides examiners with the ability to acquire, examine and generate customized reports including data such as the Card Identity (ICCID), Stored Dialing Numbers (ADN), Fixed Dialing numbers (FDN), Subscriber Number (MSISDN), Last Numbers Dialed (LND), SMS and EMS Messages, Subscriber Identity (IMSI), Ciphering Key (Kc) and Location Information (LOCI).

Acquisition Stage

Once proper connectivity is established with the SIM, the examiner is prompted for the correct PIN, if the SIM is protected, before acquisition begins. If the SIM does not contain a PIN acquisition begins by clicking OK as illustrated below in Figure 51.

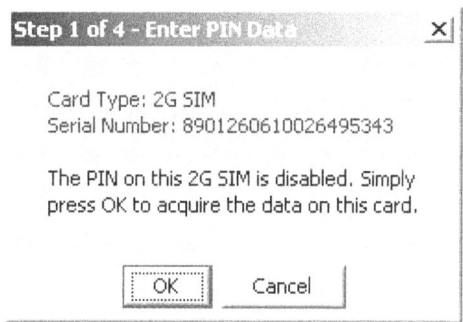

Figure 51: Acquisition Wizard

After a successful acquisition, the entire SIM contents can be saved and stored in USIMdetective's proprietary format for later processing. USIMdetective uses an internal hashing facility to ensure the integrity of case data and detect whether tampering occurred during storage. USIMdetective uses SHA1 and MD5 integrity checks to ensure that the original file is consistent with the reopened case file. Figure 52 provides a snapshot of the User Interface, which provides the examiner with a self-explanatory interface to supported functions.

---

[19] Additional information on USIMdetective can be found at: **http://www.quantaq.com/**

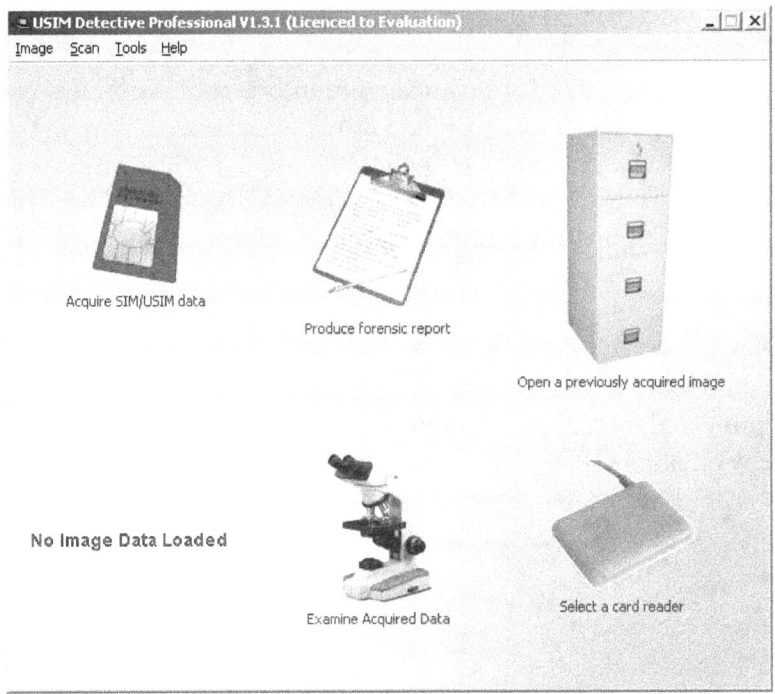

Figure 52: User Interface

Search Functionality

USIMdetective does not have a built-in search engine or search facilities. However, data can be viewed manually by browsing through the tree structure and selecting individual data items or the html generated reported can be searched using a third-party tool.

Graphics Library

USIMdetective does not display graphic files of any type, although the raw data associated with small (16x16) and large (32x32) pixel images is present. Typically, these types of images are embedded in an MMS/EMS message, and are reported alongside the textual content of the SMS/EMS message.

Report Generation

Report generation begins by prompting the examiner with case-specific details such as investigator name, date/time, case id, evidence number, and notes specific to the investigation as illustrated below in Figure 53.

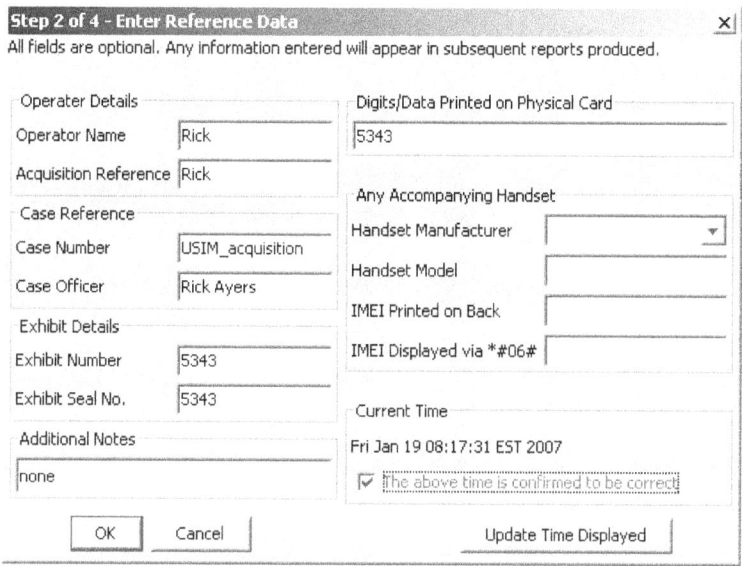

Figure 53: Acquisition Notes

The final report can be exported in multiple formats as illustrated below in Figure 54. The examiner is provided with the option to include or exclude specific data elements associated with a report type.

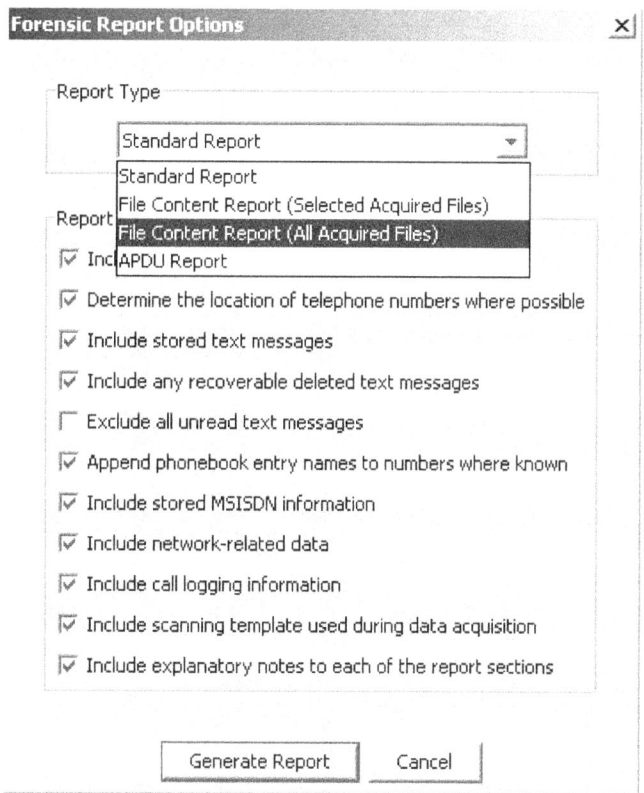

Figure 54: Report Generation

An excerpt of the report is illustrated below in Figure 55 and Figure 56. The report includes information relevant to the case as mentioned earlier and data items selected by the examiner that are relative to the case or incident.

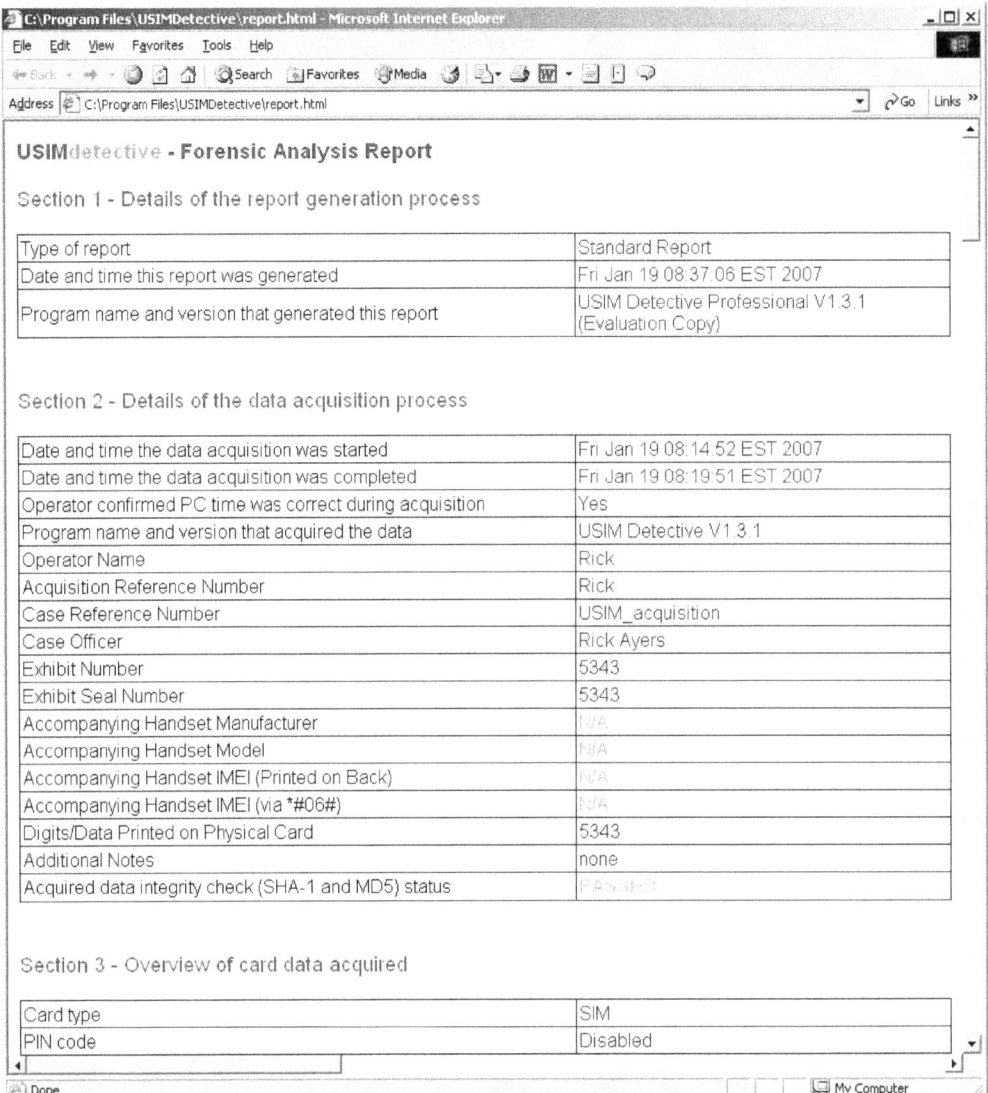

Figure 55: Report Excerpt

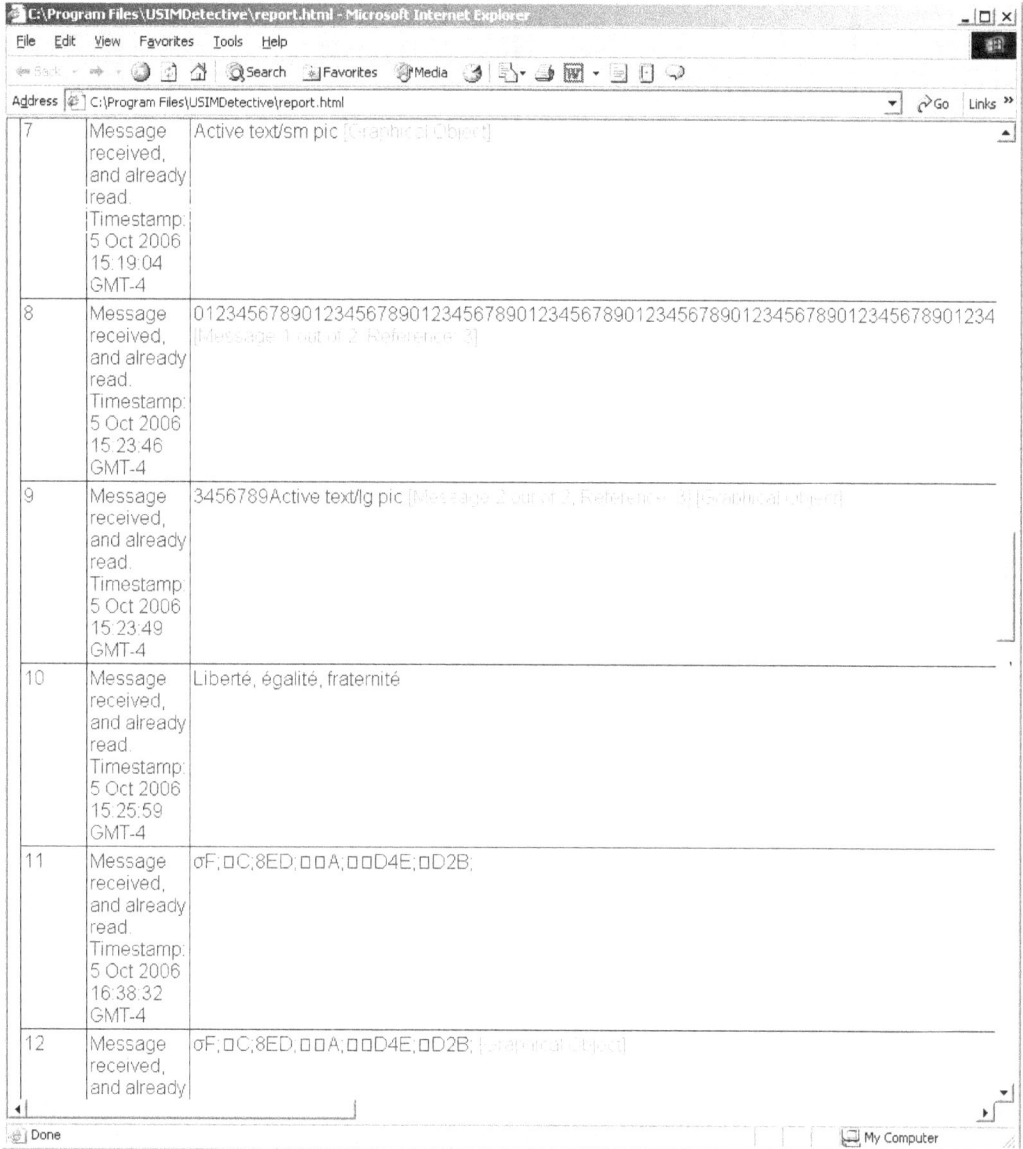

Figure 56: Report Excerpt

Scenario Results

Table 31 summarizes the results from applying the scenarios listed at the left of the table to the SIMs across the top. More information can be found in Appendix L: USIMdetective – External SIM Results.

Table 31: SIM Card Results Matrix - External Reader

Scenario	SIM		
	5343	8778	1144
Basic Data	Meet	Meet	Meet
Location Data	Meet	Meet	Meet

Scenario	SIM		
	5343	8778	1144
EMS Data	Below	Below	Below
Foreign Language Data	Below	Below	Below

Conclusions

Forensic examination of cellular devices is a growing subject area in computer forensics. Consequently, cell phone forensic tools are a relatively recent development and in the early stages of maturity. Forensic examination tools translate data to a format and structure that is understandable by the examiner and can be effectively used to identify and recover evidence. However, tools may contain some degree of inaccuracies. For example, the implementation of a tool may contain a programming error; a specification used by the tool to translate encoded bits into data comprehensible by the examiner may be inaccurate or out of date; or the protocol structure generated by the cellular device as input may be incorrect, causing the tool to function improperly. Over time, experience with a tool provides an understanding of its limitations, allowing an examiner to compensate where possible for any shortcomings or to turn to other means of recovery.

While the tools discussed in this paper generally performed well and have adequate functionality, new versions are expected to improve and better meet investigative requirements. This has been the case overall for those tools summarized previously in *NISTIR 7250 Cell Phone Forensic Tools: An Overview and Analysis*. No single forensic tool, however, addresses the full range of devices that may be encountered in practice. To cover the broadest range of mobile phones and (U)SIMs, examiners should be expected to have a toolkit that contains several well-chosen forensic tools.

The following criteria highlight some items to consider when choosing among available tools:
- Usability – the ability to present data in a form that is useful to an investigator.
- Comprehensive – the ability to present all recoverable data present on the device to an investigator so that evidence pertaining to an investigation can be identified.
- Accuracy – the quality that the output of the tool has been verified and a margin of error ascertained.
- Deterministic – the ability for the tool to produce the same output when given the same set of instructions and input data.
- Verifiable – the ability to ensure accuracy of the output by having access to intermediate translation and presentation results.
- Acceptance – the degree of peer review and agreement about the methodology or technique used by the tool.
- Quality – the technical support, reliability, and maintenance provided by the manufacturer
- Capability – the supported devices, feature set, performance, and richness of features with regard to flexibility and customization
- Affordability – the cost versus the associated benefits in productivity

Glossary of Acronyms

ADN (Abbreviated Dialing Numbers) – phone book entries kept on the SIM.

CDMA (Code Division Multiple Access) – a spread spectrum technology for cellular networks based on the Interim Standard-95 (IS-95) from the Telecommunications Industry Association (TIA).

EDGE (Enhanced Data for GSM Evolution) – an upgrade to GPRS to provide higher data rates by joining multiple time slots.

EMS (Enhanced Messaging Service) – an improved message system for GSM mobile phones allowing picture, sound, animation and text elements to be conveyed through one or more concatenated SMS messages.

ESN (Electronic Serial Number) – a unique 32-bit number programmed into CDMA phones when they are manufactured.

FCC ID (Federal Communications Commission identification number) – an identifier found on all wireless phones legally sold in the US, which is issued by the FCC.

FDN (Fixed Dialing Numbers) – a set of phone numbers kept on the SIM that the phone can call exclusively of any others (i.e., all other numbers are disallowed).

FPLMN (Forbidden PLMNs) – a list of Public Land Mobile Networks (PLMNs) maintained on the SIM that the phone cannot automatically contact, usually because service was declined by a foreign provider.

GID1 (Group Identifier Level 1) – an identifier for a particular SIM and handset association, which can be used to identify a group of SIMs involved in a particular application.

GID2 (Group Identifier Level 2) – a GID1-like identifier.

GPRS (General Packet Radio Service) – a packet switching enhancement to GSM and TDMA wireless networks to increase data transmission speeds.

GRPSLOCI (GPRS Location Information) – the Routing Area Information (RAI), Routing Area update status, and other location information maintained on the SIM.

GSM (Global System for Mobile Communications) – a set of standards for second generation cellular networks currently maintained by the 3rd Generation Partnership Project (3GPP).

HTTP (HyperText Transfer Protocol) – a standard method for communication between clients and Web servers.

ICCID (Integrated Circuit Card Identification) – a unique and immutable identifier maintained within the SIM.

iDEN (Integrated Digital Enhanced Network) – a proprietary mobile communications technology developed by Motorola that combine the capabilities of a digital cellular telephone with two-way radio.

IM (Instant Messaging) – a facility for exchanging messages in real-time with other people over the Internet and tracking the progress of the conversation.

IMEI (International Mobile Equipment Identity) – a unique number programmed into GSM and UMTS mobile phones.

IMSI (International Mobile Subscriber Identity) – a unique number associated with every GSM mobile phone user.

IMAP (Internet Message Access Protocol) – a method of communication used to read electronic mail stored in a remote server.

LND (Last Numbers Dialed) – a log of last numbers dialed, similar to that kept on the phone, but kept on the SIM without a timestamp.

LOCI (Location Information) – the Location Area Identifier (LAI) of the phone's current location, continuously maintained on the SIM when the phone is active and saved whenever the phone is turned off.

MMS (Multimedia Messaging Service) – an accepted standard for messaging that lets users send and receive messages formatted with text, graphics, photographs, audio, and video clips.

MSISDN (Mobile Subscriber Integrated Services Digital Network) – the international telephone number assigned to a cellular subscriber.

PIM (Personal Information Management) – data types such as contacts, calendar entries, tasks, notes, memos and email that may be synchronized from PC to device and vice-versa.

POP (Post Office Protocol) – a standard protocol used to receive electronic mail from a server.

SIM (Subscriber Identity Module) – a smart card chip specialized for use in GSM equipment.

SMS (Short Message Service) – a mobile phone network facility that allows users to send and receive alphanumeric text messages of up to 160 characters on their cell phone or other handheld device

SMS (Short Message Service) Chat – a facility for exchanging messages between mobile phone users in real-time via SMS text messaging, which allows previous messages from the same conversation to be viewed.

SMTP (Simple Mail Transfer Protocol) – the primary protocol used to transfer electronic mail messages on the Internet.

UMTS (Universal Mobile Telecommunications System) – a third-generation (3G) mobile phone technologies standardized by the 3GPP as the successor to GSM.

USIM (UMTS Subscriber Identity Module) – a module similar to the SIM in GSM/GPRS networks, but with additional capabilities suited to 3G networks.

WAP (Wireless Application Protocol) – a standard that defines the way in which Internet communications and other advanced services are provided on wireless mobile devices.

WIM (WAP Identity Module) – a security module implemented in the SIM that provides a trusted environment for using WAP related applications and services on a mobile device via a WAP gateway.

WiFi (Wireless Fidelity) – a generic term that refers to a wireless local area network that observes the IEEE 802.11 protocol.

WML (Wireless Markup Language) – a stripped down version of HTML to allow mobile devices to access Web sites and pages that have been converted from HTML to the more basic text format supported.

XHTML (eXtensible HyperText Markup Language) – a unifying standard that brings the XML benefits of easy validation and troubleshooting to HTML.

XML (Extensible Markup Language) – a flexible text format designed to describe data for electronic publishing.

Appendix A: Device Seizure Results – Smart Devices

The scenarios were performed on a Forensic Recovery of Evidence Device (FRED) running Windows XP SP2. Device Seizure version 1.1 was used to acquire data from Palm OS, Pocket PC and BlackBerry devices with cell phone capabilities.

Blackberry 7750

The following scenarios were executed on a Verizon CDMA BlackBerry 7750 Java-based wireless handheld device running v3.7.1.36 (Platform 1.4.0.37).

Connectivity and Retrieval: The device contents were successfully acquired. If authentication mechanisms are applied, the proper pass-phrase must be provided within 10 attempts. The correct pass-phrase must be provided for both Memory and Database acquisitions. Basic subscriber and service provider information (e.g., ESN, MSID, FCCID) were not found. The reported memory size is inconsistent with the total size found on the device via the Options->Status screen. (Meet)

PIM Applications: All PIM data (i.e., Address Book, Calendar, Tasks, Memos) was found in the corresponding databases (i.e., Address Book, Calendar, Tasks, Memos database) and reported. All deleted PIM data was found and reported in the memory window. (Meet)

Dialed/Received Phone Calls: All dialed/received phone calls were found and reported in the Phone Call Log, the Phone Hotlist database, and the memory window when conducting a search. Deleted calls were found in the memory window. (Meet)

SMS/MMS Messaging: All active incoming and outgoing SMS messages were found and reported in the Messages folder and the memory window. All deleted SMS messages were found and reported in the memory window. MMS messages are not supported. (Meet)

Internet Messaging: All data content associated with sent and received email messages was found and reported in the Messages database and the memory window. Deleted messages were found and reported in the memory window. The Blackberry device does not support Instant Messaging; therefore, this part of the scenario does not apply. (Meet)

Web Applications: N.A. – In order for the Verizon BlackBerry 7750 to utilize the Web viewer, Verizon customers must purchase a third party solution such as MobileWeb4U Mobile Web WAP Gateway. Verizon does not provide a public WAP gateway with this capability. (NA)

Text File Formats: Data content associated with text files (i.e., .txt, .doc, .pdf) was found only for .txt files when sent via email or the BlackBerry desktop manager protocol. No deleted data was found. (Below)

Graphics File Formats: Graphics file (e.g., .bmp, .jpg, .gif, .png, .tif) data content was not found or displayed when sent via email. The BlackBerry desktop manager protocol does not allow for transferring image files to the device. The filename and subject line of the email were found and reported. (Miss)

Compressed File Archive Formats: Compressed data file (i.e., `.zip`, `.rar`, `.exe`, `.tgz`) content was not found. (Miss)

Misnamed Files: Misnamed file (e.g., `.txt` file renamed with a `.dll` extension) data content was not found. (Miss)

Peripheral Memory Cards: N.A. – The BlackBerry 7750 does not allow for removable media. (NA)

Acquisition Consistency: Two consecutive acquisition produce different overall hashes of the memory. However, the individual database files acquired were consistent. (Meet)

Cleared Devices: Two approaches exist for performing a hard reset on a BlackBerry device: 1) not supplying the correct pass-phrase within 10 attempts 2) accessing the BlackBerry Desktop Manager and clearing all databases via the Backup/Restore advanced menu. The first approach forces the user to re-download the OS before the device is functional. No data was found. (Meet)

The following data was found after clearing the device through the desktop manager: PIM data (i.e., Address Book, Calendar), dialed/received phone calls and Internet Messaging data (i.e., subject/filename). (Above)

Power Loss: The BlackBerry 7750 was repopulated with the above scenarios, then completely drained of all battery power and reacquired. All data was found as reported above. (Above)

BlackBerry 7780

The following scenarios were executed on an AT&T GSM BlackBerry 7780 Java-based wireless handheld device running v3.7.1.59 (Platform 1.6.1.48).

Connectivity and Retrieval: The device contents were successfully acquired with or without the SIM present. If authentication mechanisms are applied, the proper pass-phrase must be provided within 10 attempts. The correct pass-phrase must be provided for both Memory and Database acquisitions. Basic subscriber and service provider information (e.g., IMEI, ICCID, MSISDN) were not found. The reported memory size is inconsistent with the total size found on the device via the Options->Status screen. (Meet)

PIM Applications: All PIM data was found and reported (i.e., Address Book, Calendar, Tasks, Memos) in the corresponding databases (i.e., Address Book, Calendar, Tasks, Memos database). All deleted PIM data was found and reported in the memory window. (Meet)

Dialed/Received Phone Calls: All dialed/received phone calls were found and reported in the Phone Call Log, Phone Hotlist database, and the memory window when conducting a search. Deleted calls were found in the memory window. (Meet)

SMS/MMS Messaging: All active incoming and outgoing SMS messages were found and reported in the Messages folder and the memory window when conducting a search on message content. All deleted SMS messages were found and reported in the memory window. MMS messages are not supported. (Meet)

Internet Messaging: All data content associated with sent and received email messages was found and reported in the Messages database and the memory window. Deleted messages were found and reported in the memory window. The BlackBerry device does not support Instant Messaging and therefore omitted from this scenario. (Meet)

Web Applications: Visited URLs and search engine queries were found and reported. Textual Web content pertaining to URLs was not found. No graphical images of visited sites were found and displayed. (Below)

Text File Formats: Data content associated with text files (i.e., .txt, .doc, .pdf) was found only for .txt files when sent via email or the BlackBerry desktop manager protocol. The filename and subject line of the email were found and reported for all text-based files. No deleted data was found. (Below)

Graphics File Formats: Graphics file (e.g., .bmp, .jpg, .gif, .png, .tif) data content was not found or displayed when sent via email. The BlackBerry desktop manager protocol does not allow for transferring image files to the device. The filename and subject line of the email were found and reported. (Miss)

Compressed File Archive Formats: Compressed data file (i.e., .zip, .rar, .exe, .tgz) content was not found. (Miss)

Misnamed Files: Misnamed file (e.g., .txt file renamed with a .dll extension) data content was not found. (Miss)

Peripheral Memory Cards: N.A. – The BlackBerry 7780 does not allow for removable media. (NA)

Acquisition Consistency: Two consecutive acquisition produce different overall hashes of the memory. The individual database files acquired were consistent. (Meet)

Cleared Devices: Two approaches exist for performing a hard reset on a BlackBerry device: 1) not supplying the correct pass-phrase within 10 attempts, and 2) accessing the BlackBerry Desktop Manager and clearing all databases via the Backup/Restore advanced menu. The first approach forces the user to re-download the OS before the device is functional. No data was found. (Meet)

The following data was found after clearing the device through the desktop manager: PIM data (i.e., Address Book, Calendar), dialed/received phone calls and Internet Messaging data (i.e., subject/filename). (Above)

Power Loss: The BlackBerry 7780 was repopulated with the above scenarios, then completely drained of all battery power and reacquired. All data was found as reported above. (Above)

Kyocera 7135

The following scenarios were conducted on a Verizon Kyocera 7135 running Palm OS version 4.1.

Connectivity and Retrieval: The device contents were successfully acquired. If authentication mechanisms are applied, the proper pass-phrase has to be provided in order to begin acquisition. Basic subscriber and service provider information (e.g., ESN, MSID, FCCID) were not found. Network carrier and the phone number was found in the NetworkDB file and reported. Since internal memory could not be acquired, memory size was not reported and could not be verified. (Meet)

PIM Applications: All PIM data was found (i.e., Address Book, Calendar, Tasks, Memos) in the corresponding database/folder (i.e., AddressDB, DatebookDB, ToDoDB, MemoDB files) and reported. The following deleted PIM data was found and reported in association to the Address Book entries: Company Name, found in the AddressCompaniesDB, and Titles, found in the AddressTitlesDB. (Meet)

Dialed/Received Phone Calls: All dialed/received phone calls were found and reported in the kwc_CallHistoryDB file. Deleted phone calls were found in the memory dump. (Meet)

SMS/MMS Messaging: All active incoming and outgoing SMS messages were found in the kwc_messages file and reported. Deleted SMS messages were not found or reported. MMS Messaging is not supported. (Below)

Internet Messaging: All data content associated with sent and received email messages was found in the MailDB, pdQmailMsgs and the MMPROIII Message files and reported. Deleted messages were not found. (Below)

Web Applications: Visited URLs were found and reported, while search queries performed were not found. Textual or graphical Web content pertaining to URLs was not found. (Below)

Text File Formats: Data content associated with text files (i.e., .txt, .doc, .pdf) was found and reported for .txt files. However, .pdf and .doc file content sent via email and Hotsync was not found in a readable format. Necessary software that allows each of the above file types to be read was installed on the device. Deleted text file data was found and reported for .txt files. (Meet)

Graphics File Formats: Graphic file (i.e., .bmp, .jpg, .gif, .png, .tif) data was not found when sent via email or Hotsync. (Miss)

Compressed File Archive Formats: Compressed data file (i.e., .zip, .rar, .exe, .tgz) content was not found. (Miss)

Misnamed Files: Misnamed file (e.g., `.txt` file renamed with a `.dll` extension) data content was found and reported for text-based files when sent via email. Unknown file types are not accepted by the Hotsync protocol. (Meet)

Peripheral Memory Cards: No data residing on a 128 MB MMC populated with various files (i.e., text, graphics, audio, compressed archive files, misnamed files) was found. (Miss)

Acquisition Consistency: Two consecutive acquisition produce different hashes on the following database files: NetworkDB and Saved Preferences. (Below)

Cleared Devices: A Hard Reset was performed by holding the reset button while pressing the illumination key. No data was found. (Meet)

Power Loss: The Kyocera 7135 was repopulated with the above scenarios, then completely drained of all battery power and reacquired. No data was found. (Meet)

Motorola MPx220

The following scenarios were conducted on a Cingular GSM Motorola MPx220 running Microsoft Windows Mobile 2004 for Pocket PC Phone Edition.

Connectivity and Retrieval: The data contents of the device were successfully acquired with or without the SIM present. If internal memory authentication mechanisms are applied, the proper pass-phrase must be provided in order for the ActiveSync connection to take place. Basic subscriber and service provider information (e.g., IMEI, ICCID, MSISDN) were not found. Network carrier and the phone number was found in the \Windows directory and reported. The reported memory size is consistent with the total memory size of the device. (Meet)

PIM Applications: PIM data was not found (i.e., Address Book, Calendar, Tasks, Memos). (Miss)

Dialed/Received Phone Calls: Dialed/received phone calls were not found. (Miss)

SMS/MMS Messaging: SMS messages were not found. Textual content of all active incoming MMS messages were found and reported. MMS attachments were not found. Deleted MMS messages were not found. (Below)

Internet Messaging: All data content associated with sent and received email messages was found and reported. Deleted messages were not found. (Below)

Web Applications: Visited URLs and search engine queries were found and reported. Textual Web content pertaining to visited URLs was not found. No graphical images of visited sites were found. (Below)

Text File Formats: Data content associated with text files (i.e., `.txt`, `.doc`, `.pdf`) was found and reported. Deleted text files were not found. (Below)

Graphics File Formats: Graphics file (i.e., .bmp, .jpg, .gif, .png, .tif) data content was found and reported. .png files were found but not displayed in the graphics library. Deleted graphics files were not found. (Below)

Compressed File Archive Formats: Compressed data file (i.e., .zip, .rar, .exe, .tgz) content was found and reported. (Meet)

Misnamed Files: Misnamed file (e.g., .txt file renamed with a .dll extension) data content was found and reported. (Meet)

Peripheral Memory Cards: Data residing on a 256 MB Mini SD Card populated with various files (i.e., text, graphics, audio, compressed archive files, misnamed files) was found and reported. Deleted files were not found. (Below)

Acquisition Consistency: Two consecutive acquisition produce different overall hashes of the memory. The individual files acquired were consistent. (Meet)

Cleared Devices: A Hard Reset was performed by holding down the action button and pressing the power button. No data was found. (Meet)

Power Loss: The Motorola MPx220 was repopulated with the above scenarios, then completely drained of all battery power and reacquired. The following data was found: MMS messages (text content) and Internet Messaging data (i.e., sent/received email). Individual files stored in the /storage directory were found and reported. (Above)

Samsung i70

The following scenarios were conducted on a Verizon CDMA Samsung i700 running Microsoft Windows Mobile 2004 for Pocket PC Phone Edition.

Connectivity and Retrieval: The device contents were successfully acquired. If authentication mechanisms are applied, the proper pass-phrase must be provided in order for the ActiveSync connection to take place. Basic subscriber and service provider information (e.g., ESN, MSID, FCCID) were not found. Network carrier and the phone number was found in the \Windows directory and reported. The reported memory size is consistent with the total memory size of the device. (Meet)

PIM Applications: All PIM data was found (i.e., Address Book, Calendar, Tasks, Memos) in the corresponding database/folder (i.e., Contacts, Appointments, Tasks database, Memos) and reported. Deleted PIM data was found and reported. (Meet)

Dialed/Received Phone Calls: All dialed/received phone calls were found in the clog database and the memory image and reported. Phone numbers were found by issuing the following search pattern: (.a.a.a)..p.p.p.-.s.s.s.s. Deleted calls were found in the memory image. (Meet)

SMS/MMS Messaging: All active incoming and outgoing SMS messages were found in the \Windows\Messaging folder and the fldr100171c file and reported. Deleted SMS messages were not found. MMS messaging is not supported. (Below)

Internet Messaging: All data content associated with sent and received email messages was found in the \Windows\Messaging file and reported. Deleted message file content (e.g., subject, body text) was not found. (Below)

Web Applications: Visited URLs and search engine queries were found and reported. Textual Web content pertaining to URLs was found and reported. No graphical images of visited sites were found or displayed. (Below)

Text File Formats: Data content associated with text files (i.e., .txt, .doc, .pdf) was found and reported when transferred via email and ActiveSync. Deleted text file data was not found. (Below)

Graphics File Formats: Graphic file (i.e., .bmp, .jpg, .gif, .png, .tif) filenames and data content were found and displayed when sent via email and ActiveSync. Deleted graphic file data was not found. (Below)

Compressed File Archive Formats: Compressed data (i.e., .zip, .rar, .exe, .tgz) filenames and content were found and reported when sent via email and ActiveSync. (Meet)

Misnamed Files: Misnamed file (e.g., .txt file renamed with a .dll extension) data content was found when sent via email and ActiveSync. (Meet)

Peripheral Memory Cards: Data residing on a 128 MB MMC populated with various files (i.e., text, graphics, audio, compressed archive files, misnamed files) was found and reported. Deleted files were not found. (Below)

Acquisition Consistency: Two consecutive acquisition produce different hashes on the following database files: \Categories Database, \ConfigMetabase, SchedSync.dat, DB_notify_events, DB_notify_queue, pmailFolders, pmailMsgClasses, pmailNamedProps, pmailServices, and Speed.db. (Below)

Cleared Devices: A Hard Reset was performed by holding down the power button while pressing the reset button with the stylus, then releasing the power button. No data was found. (Meet)

Power Loss: The Samsung i700 Pocket PC was repopulated with the above scenarios and completely drained of all battery power and reacquired. No data was found. (Meet)

PalmOne Treo 6

The following scenarios were conducted on the PalmOne Treo 600 running Palm OS version 5.2.1.

Connectivity and Retrieval: The device contents were successfully acquired with or without the SIM present. When authentication mechanisms were applied, the proper pass-phrase had to be provided in order to begin acquisition. Basic subscriber and service provider information (e.g., IMEI, ICCID, MSISDN) were not found. The memory size was not reported and could not be verified. (Meet)

PIM Applications: All active PIM data was found (i.e., Address Book, Calendar, Tasks, Memos) in the corresponding database/folder (i.e., AddressDB, DatebookDB, ToDoDB, MemoDB files) and reported. Deleted PIM data was found in the memory file. (Meet)

Dialed/Received Phone Calls: All dialed/received phone calls were found in the PhoneCallDB file and reported. Deleted phone calls were found in the memory file. (Below)

SMS/MMS Messaging: All active SMS messages were found and reported in the Msg Database.pdb file. MMS messages and corresponding filenames were found and reported. However, the associated multimedia files were not viewable. Deleted SMS/MMS messages were found in the memory file. (Below)

Internet Messaging: All data content associated with sent and received email messages was found in the Email_libr_HsMp_BDC79AAB file and reported. Deleted message file content (e.g., subject, body text) was not found, but the email address was found in the EmailAddressDB.pdb file. (Below)

Web Applications: Visited URLs and search engine queries were found and reported. Textual Web content pertaining to URLs was found and reported. No graphical images of visited sites were found or displayed. (Below)

Text File Formats: Data content associated with text files (i.e., `.txt`, `.doc`, `.pdf`) was found and reported for `.doc` and `.txt` files. However, `.pdf` file content data sent via email and Hotsync was not found in a readable format. Necessary software that allows each of the above file types to be read was installed on the device. Deleted text file data was found and reported for `.doc` and `.txt` files. (Meet)

Graphics File Formats: Graphic file (i.e., `.bmp`, `.jpg`, `.gif`, `.png`, `.tif`) data was found for file types supported by Palm OS. (Below)

Compressed File Archive Formats: Compressed data file (i.e., `.zip`, `.rar`, `.exe`, `.tgz`) content was not found, but the filename embedded in the compressed file was found and reported when sent via email. (Below)

Misnamed Files: Misnamed file (e.g., `.txt` file renamed with a `.dll` extension) data content was found and reported for text-based files, but only when sent via email. Unknown file types are not accepted by the Hotsync protocol. (Meet)

Peripheral Memory Cards: No data residing on a 128 MB MMC populated with various files (i.e., text, graphics, audio, compressed archive files, misnamed files) was found or reported. (Miss)

Acquisition Consistency: Individual files and database that have not been modified maintain consistent hashes between consecutive acquisitions. (Meet)

Cleared Devices: A Hard Reset was performed by holding down the reset button and pressing the power key. No data was found. (Meet)

Power Loss: The Treo 600 was repopulated with the above scenarios, then completely drained of all battery power and reacquired. No data was found. (Meet)

Appendix B: Device Seizure Results – Cell Phones

The scenarios were performed on a Forensic Recovery of Evidence Device (FRED) running Windows XP SP2. Device Seizure version 1.1 was used to acquire data from the following cell phones: Audiovox 8910, Ericsson T68i, LG 4015, Motorola C333, Motorola V66, Motorola v300, Nokia 3390, Nokia 6610i and the Sanyo PM8200.

Audiovox 8910

The following scenarios were conducted on a pre-paid CMDA Audiovox 8910. Device Seizure connectivity was established by selecting LG (CDMA).

Connectivity and Retrieval: Data-Pilot's Susteen data cable for Audiovox phones was used in order to establish connectivity with Device Seizure. The password-protected device contents were successfully acquired. Basic subscriber and service provider information was found and reported. Memory size is not reported. (Meet)

PIM Applications: All active and remnants of deleted PIM data were found and reported in the Filesystem section. (Meet)

Dialed/Received Phone Calls: All active and deleted dialed/received phone calls were found and reported in the Filesystem section. (Meet)

SMS/MMS Messaging: All active incoming and outgoing SMS/MMS messages were found and reported in the Filesystem section. Deleted messages were not found. (Below)

Internet Messaging: N.A. – The Audiovox 8910 does not support email. (NA)

Web Applications: Visited URLs were found and reported. Search engine queries, textual Web content pertaining to visited URLs and graphical images of visited sites, were not found. (Below)

Text File Formats: N.A. – The Audiovox 8910 does not support text files (e.g., .txt, .doc, .pdf). (NA)

Graphics Files Format: A connection could not be established allowing the transfer of graphic files (i.e., .bmp, .jpg, .gif .png, .tif) to the Audiovox 8910. Images were created via the internal camera. Graphic files present on the device were found and reported in the Filesystem section. Images were exported and viewed with a third party application. Deleted graphic files were not found. (Below)

Compressed File Archive Formats: N.A. – The Audiovox 8910 does not support compressed archive files (e.g., .zip, .rar, .exe, .tgz). (NA)

Misnamed Files: N.A. – The Audiovox 8910 does not support misnamed files (e.g., .txt file renamed with a .dll extension). (NA)

Peripheral Memory Cards: N.A. – The Audiovox 8910 does not allow for removable media. (NA)

Acquisition Consistency: All hashes of individual folders were consistent. (Meet)

Cleared Devices: N.A. – A Hard Reset function is not provided by the phone. (NA)

Power Loss: The Audiovox 8910 was repopulated with the above scenarios, then completely drained of all battery power and reacquired. All data was found as reported above. (Above)

Ericsson T68i

The following scenarios were conducted on a Sony Ericsson T68i. Device Seizure version 1.1 was used for acquisition.

Connectivity and Retrieval: Data-Pilot's Susteen data cable for Sony Ericsson phones was used in order to establish connectivity with Device Seizure. Proper authentication had to be provided to the password-protected device and the SIM card had to be inserted before contents were successfully acquired. Basic subscriber and service provider information was found and reported (i.e., IMEI). Memory size is not reported. (Meet)

PIM Applications: All PIM data was found and reported (i.e., Address Book, Calendar, Tasks). Deleted PIM data was not found. (Below)

Dialed/Received Phone Calls: All dialed/received phone calls were found and reported in the Phone calls folder. Deleted phone calls were not found. (Below)

SMS/MMS Messaging: All active incoming and outgoing SMS messages were found in the SMS folder and reported. Deleted SMS messages were not found. MMS Messages and attachments (i.e., graphics, sound bites) were not found or reported. (Below)

Internet Messaging: Data content associated with sent and received email messages were not found. (Miss)

Web Applications: Visited URLs, search queries performed, textual Web content and graphical images of visited sites were not found. (Miss)

Text File Formats: N.A. – The Sony Ericsson T68i does not support text files (e.g., .txt, .doc, .pdf). (NA)

Graphics Files Format: Supported graphic files (i.e., .jpg, .gif) present on the device were not found. (Miss)

Compressed File Archive Formats: N.A. – The Sony Ericsson T68i does not support compressed archive files (e.g., .zip, .rar, .exe, .tgz). (NA)

Misnamed Files: N.A. – The Sony Ericsson T68i does not support misnamed files (e.g., .txt file renamed with a .dll extension). (NA)

Peripheral Memory Cards: N.A. – The Sony Ericsson T68i does not allow for removable media. (NA)

Acquisition Consistency: All hashes of individual folders were consistent. (Meet)

Cleared Devices: A Hard Reset was performed by selecting Master Reset in the settings menu. Data contained on the SIM was found and reported. (Meet)

Power Loss: The Sony Ericsson T68i was repopulated with the above scenarios, then completely drained of all battery power and reacquired. All data was found as reported above. (Above)

LG 4015

The following scenarios were conducted on a GSM LG 4015. Device Seizure version 1.1 was used for acquisition.

Connectivity and Retrieval: The password-protected device contents were successfully acquired. Basic subscriber and service provider information was not found. Memory size is not reported. (Meet)

PIM Applications: Partial PIM data was found (i.e., Address Book, Memos) in the corresponding database/folder (i.e., Phonebook, Scheduler folders) and reported. Deleted PIM data was not found. (Below)

Dialed/Received Phone Calls: Dialed/received phone calls were not found. (Miss)

SMS/MMS Messaging: All active incoming and outgoing SMS were found and reported. No other data was found. (Below)

Internet Messaging: N.A. – The LG 4015 does not support email. (NA)

Web Applications: No data was found. (Miss)

Text File Formats: N.A. – The LG 4015 does not support text files (e.g., .txt, .doc, .pdf). (NA)

Graphics Files Format: N.A. – A connection could not be established allowing the transfer of graphic files (i.e., .bmp, .jpg, .gif .png, .tif) to the LG 4015. (NA)

Compressed File Archive Formats: N.A. – The LG 4015 does not support compressed archive files (e.g., .zip, .rar, .exe, .tgz). (NA)

Misnamed Files: N.A. – The LG 4015 does not support misnamed files (e.g., .txt file renamed with a .dll extension). (NA)

Peripheral Memory Cards: N.A. – The LG 4015 does not allow for removable media. (NA)

Acquisition Consistency: All hashes of individual folders were consistent. (Meet)

Cleared Devices: A Hard Reset was performed by selecting the Reset option in the security menu. All active data was recovered. (Above)

Power Loss: The Sanyo PM-8200 was repopulated with the above scenarios, then completely drained of all battery power and reacquired. All data was found as reported above. (Above)

Motorola C333

The following scenarios were conducted on an unlocked GSM Motorola C333. Device Seizure version 1.1 was used for acquisition.

Connectivity and Retrieval: Proper authentication had to be provided to the password-protected device and the SIM card had to be inserted before contents were successfully acquired. Basic subscriber and service provider information (i.e., IMEI) was found. Memory size is not reported. (Meet)

PIM Applications: All PIM data was found (i.e., Address Book, Calendar) in the corresponding database/folder (i.e., Phonebook, Datebook folders) and reported. Deleted PIM data was found and reported in the Filesystem folder. (Meet)

Dialed/Received Phone Calls: All dialed/received phone calls were found and reported in the Call History folder. Deleted phone calls were not found. (Below)

SMS/MMS Messaging: All active incoming and outgoing SMS messages were found in the SMS Message folder and reported. Deleted incoming and outgoing SMS messages were found in the SMS and Quick Notes dump folder. Incoming and outgoing MMS messages were found but no associated multi-media files. (Meet)

Internet Messaging: N.A. – The Motorola C333 does not support email. (NA)

Web Applications: N.A. – Internet connectivity was unable to be established. (NA)

Text File Formats: N.A. – The Motorola C333 does not support text files (e.g., `.txt`, `.doc`, `.pdf`). (NA)

Graphics Files Format: N.A. – The Motorola C333 does not support graphic files (e.g., `.bmp`, `.jpg`, `.gif`, `.png`, `.tif`). (NA)

Compressed File Archive Formats: N.A. – The Motorola C333 does not support compressed archive files (e.g., `.zip`, `.rar`, `.exe`, `.tgz`). (NA)

Misnamed Files: N.A. – The Motorola C333 does not support misnamed files (e.g., `.txt` file renamed with a `.dll` extension). (NA)

Peripheral Memory Cards: N.A. – The Motorola C333 does not allow for removable media. (NA)

Acquisition Consistency: All hashes of individual folders were consistent. (Meet)

Cleared Devices: A Hard Reset was performed by selecting Master Clear and Master Reset in the settings menu. SMS messages and Memos were found and reported in the SMS and Quick notes dump folder. (Above)

Power Loss: The Motorola C333 was repopulated with the above scenarios, then completely drained of all battery power and reacquired. All data was found as reported above. (Above)

Motorola V6

The following scenarios were conducted on an unlocked GSM Motorola V.series 66. Device Seizure version 1.1 was used for acquisition.

Connectivity and Retrieval: The password-protected device contents were successfully acquired with or without the SIM present without having to provide proper authentication. Basic subscriber and service provider information (i.e., IMEI, ICCID, IMSI) was found. Memory size is not reported. (Meet)

PIM Applications: All PIM data was found (i.e., Address Book, Calendar) in the corresponding database/folder (i.e., Phonebook, Datebook folders) and reported. Deleted PIM data was not found. (Below)

Dialed/Received Phone Calls: All dialed/received phone calls were found and reported in the Call History folder. Deleted phone calls were not found. (Below)

SMS/MMS Messaging: All active incoming and outgoing SMS messages were found in the SMS Message folder and reported. Deleted SMS messages were not found. MMS Messaging is not supported. (Below)

Internet Messaging: N.A. – The Motorola V66 does not support email. (NA)

Web Applications: N.A. – Internet connectivity was unable to be established. (NA)

Text File Formats: N.A. – The Motorola V66 does not support text files (e.g., `.txt`, `.doc`, `.pdf`). (NA)

Graphics Files Format: N.A. – The Motorola V66 does not support graphic files (e.g., `.bmp`, `.jpg`, `.gif`, `.png`, `.tif`). (NA)

Compressed File Archive Formats: N.A. – The Motorola V66 does not support compressed archive files (e.g., `.zip`, `.rar`, `.exe`, `.tgz`). (NA)

Misnamed Files: N.A. – The Motorola V66 does not support misnamed files (e.g., `.txt` file renamed with a `.dll` extension). (NA)

Peripheral Memory Cards: N.A. – The Motorola V66 does not allow for removable media. (NA)

Acquisition Consistency: All hashes of individual folders were consistent. (Meet)

Cleared Devices: A Hard Reset was performed by selecting, Master Clear/Reset in the settings menu. Deleted SMS and PIM data was found and reported. (Above)

Power Loss: The Motorola V66 was repopulated with the above scenarios, then completely drained of all battery power and reacquired. All data was found as reported above. (Above)

Motorola V30

The following scenarios were conducted on a Pay-As-You-Go GSM Motorola V300. Device Seizure version 1.1 was used for acquisition.

Connectivity and Retrieval: Proper authentication had to be provided to the password-protected device and the SIM card had to be inserted before contents were successfully acquired. Basic subscriber and service provider information was found and reported (i.e., IMEI, IMSI). Memory size is not reported. (Meet)

PIM Applications: All PIM data was found (i.e., Address Book, Calendar) in the corresponding database/folder (i.e., Contacts, Calendar folders) and reported. Deleted PIM data was not found. (Below)

Dialed/Received Phone Calls: All dialed/received phone calls were found and reported in the Calls folder. Deleted phone calls were not found. (Below)

SMS/MMS Messaging: All active incoming and outgoing SMS messages were found in the SMS Message folder and reported. Deleted outgoing SMS messages were found in the SMS and Quick Notes dump folder. MMS messages and attachments (i.e., graphics, sound bytes) were found and reported. (Meet)

Internet Messaging: Incoming/outgoing emails were found and reported. Deleted messages were found and reported in the raw memory dump. (Meet)

Web Applications: Visited URLs, search queries performed, textual Web content and graphical images of visited sites were not found. (Miss)

Text File Formats: Data content associated with text files (i.e., `.txt`, `.doc`, `.pdf`) was not found when sent via email. The filename and subject line of the email were found and reported. (Miss)

Graphics Files Format: A connection could not be established allowing the transfer of graphic files (i.e., .bmp, .jpg, .gif .png, .tif) to the Motorola V300. Images were created via the internal camera. Graphics files were found and reported. (Meet)

Compressed File Archive Formats: N.A. – The Motorola V300 does not support compressed archive files (e.g., .zip, .rar, .exe, .tgz). (NA)

Misnamed Files: Misnamed file (e.g., .txt file renamed with a .dll extension) data content was not found when sent via email. The filename and subject line of the email were found and reported. (Miss)

Peripheral Memory Cards: N.A. – The Motorola V300 does not allow for removable media. (NA)

Acquisition Consistency: All hashes of individual folders were consistent. (Meet)

Cleared Devices: A Hard Reset was performed by selecting the Master Clear and Master Reset options in the settings menu. SMS messages were recovered. (Above)

Power Loss: The Motorola V300 was repopulated with the above scenarios, then completely drained of all battery power and reacquired. All data was found as reported above. (Above)

Nokia 3390

The following scenarios were conducted on a GSM Nokia 3390. Device Seizure version 1.1 was used for acquisition.

Connectivity and Retrieval: The password-protected device contents were successfully acquired with or without the SIM present providing proper internal memory authentication. Basic subscriber and service provider information was found (i.e., IMEI). Memory size is not reported. (Meet)

PIM Applications: All PIM data was found and reported (i.e., Phonebook, Calendar). Deleted PIM data was not found. (Below)

Dialed/Received Phone Calls: All dialed/received phone calls were found and reported in the Phone calls folder. Deleted phone calls were not found. (Below)

SMS/MMS Messaging: All active incoming SMS messages were found in the SMS folder and reported. Outgoing SMS messages were not found. Deleted SMS messages were not found. MMS messages are not supported. (Below)

Internet Messaging: N.A. – The Nokia 3390 does not support email. (NA)

Web Applications: The Nokia 3390 does not support browsing the Web but allows for Instant Messaging. No data was found. (Miss)

Text File Formats: N.A. – The Nokia 3390 does not support text files (e.g., .txt, .doc, .pdf). (NA)

Graphics Files Format: N.A. – The Nokia 3390 does not support graphic files (e.g., .bmp, .jpg, .gif, .png, .tif). (NA)

Compressed File Archive Formats: N.A. – The Nokia 3390 does not support compressed archive files (e.g., .zip, .rar, .exe, .tgz). (NA)

Misnamed Files: N.A. – The Nokia 3390 does not support misnamed files (e.g., .txt file renamed with a .dll extension). (NA)

Peripheral Memory Cards: N.A. – The Nokia 3390 does not allow for removable media. (NA)

Acquisition Consistency: All hashes of individual folders were consistent. (Meet)

Cleared Devices: N.A. – A Hard Reset function is not provided by the phone. (NA)

Power Loss: The Nokia 3390 was repopulated with the above scenarios, then completely drained of all battery power and reacquired. All data was found as reported above. (Above)

Nokia 6610i

The following scenarios were conducted on an unlocked GSM Nokia 6610i. Device Seizure version 1.1 was used for acquisition.

Connectivity and Retrieval: Proper authentication had to be provided to the password-protected device when the SIM was present, although acquisition was successful without the SIM. Basic subscriber and service provider information was found and reported (i.e., IMEI). Memory size is not reported. (Meet)

PIM Applications: All PIM data was found and reported (i.e., Phonebook, Calendar). Deleted PIM data was not found. (Below)

Dialed/Received Phone Calls: All dialed/received phone calls were found and reported in the Call Logs folder. Deleted phone calls were not found. (Below)

SMS/MMS Messaging: All active incoming and outgoing SMS messages were found in the SMS History folder and reported. Active MMS messages and associated files (e.g., graphic, audio files) were found. Deleted SMS/MMS messages were not found. (Below)

Internet Messaging: N.A. – The Nokia 6610i does not support email. (NA)

Web Applications: Visited URLs were found and reported, while search queries performed were not found. Textual Web content pertaining to visited URLs was not found. No graphical images of visited sites were found and displayed. (Below)

Text File Formats: Data content associated with text files (i.e., .txt, .doc, .pdf) was found and reported in the Filesystem folder and could be viewed after saved to the forensic workstation. Deleted text files were not found. (Below)

Graphics Files Format: Graphic files (e.g., .bmp, .jpg, .gif, .png, .tif) present on the device were found and reported. Deleted graphic files were not found. (Below)

Compressed File Archive Formats: Compressed data file (i.e., .zip, .rar, .exe, .tgz) content was found and reported in the File System folder and could be viewed after saved to the forensic workstation. (Meet)

Misnamed Files: Misnamed files (e.g., .txt file renamed with a .dll extension) were found and reported in the File System folder and could be viewed with the proper application after saved to the forensic workstation. (Meet)

Peripheral Memory Cards: N.A. – The Nokia 6610i does not allow for removable media. (NA)

Acquisition Consistency: All hashes of individual folders were consistent. (Meet)

Cleared Devices: N.A. – A Hard Reset function is not provided by the phone. (NA)

Power Loss: The Nokia 6610i was repopulated with the above scenarios, then completely drained of all battery power and reacquired. All data was found as reported above. (Above)

Sanyo PM-820

The following scenarios were conducted on a CMDA Sanyo 8200. Device Seizure connectivity was established by selecting LG (CDMA).

Connectivity and Retrieval: The password-protected device contents were successfully acquired. Basic subscriber and service provider information was found and reported. Memory size is not reported. (Meet)

PIM Applications: All active PIM data was found and reported in the corresponding folder (e.g., Phonebook, Calendar). Tasks entries were found and reported in the Filesystem section. Deleted contact PIM data was found and reported in the Filesystem section. Data remnants of deleted Calendar and Tasks entries were found and reported in the Filesystem section. (Meet)

Dialed/Received Phone Calls: All active and deleted dialed/received phone calls were found and reported in the Filesystem section. (Meet)

SMS/MMS Messaging: All active incoming and outgoing SMS and outgoing MMS messages (textual content) were found and reported in the Filesystem section. Deleted outgoing SMS/MMS messages (textual content) were found and reported in the Filesystem section. Incoming MMS messages (i.e., textual content) were not found. (Meet)

Internet Messaging: Data content associated with sent and received email messages was not found. (Miss)

Web Applications: Visited URLs or web content pertaining to visited sites were not found. (Miss)

Text File Formats: N.A. – The Sanyo PM-8200 does not support text files (e.g., .txt, .doc, .pdf). (NA)

Graphics Files Format: A connection could not be established allowing the transfer of graphic files (i.e., .bmp, .jpg, .gif .png, .tif) to the Sanyo PM-8200. Images were created via the picture camera. Graphic files present on the device were not found. (Miss)

Compressed File Archive Formats: N.A. – The Sanyo PM-8200 does not support compressed archive files (e.g., .zip, .rar, .exe, .tgz). (NA)

Misnamed Files: N.A. – The Sanyo PM-8200 does not support misnamed files (e.g., .txt file renamed with a .dll extension). (NA)

Peripheral Memory Cards: N.A. – The Sanyo PM-8200 does not allow for removable media. (NA)

Acquisition Consistency: All hashes of individual folders were consistent. (Meet)

Cleared Devices: A Hard Reset was performed by selecting the Reset option in the security menu. All active data was recovered. (Above)

Power Loss: The Sanyo PM-8200 was repopulated with the above scenarios, then completely drained of all battery power and reacquired. All data was found as reported above. (Above)

Appendix C: GSM .XRY Results

The scenarios were performed on a Forensic Recovery of Evidence Device (FRED) running Windows XP SP2. GSM .XRY version 3.0 was used to acquire data from the following cell phones: Ericsson T68i, Motorola C333, Motorola V66, Motorola V300, Nokia 6610i, Nokia 6200, and Nokia 7610.

Ericsson T68i

The following scenarios were conducted on a Sony Ericsson T68i. GSM .XRY version 3.0 was used for acquisition.

Connectivity and Retrieval: Proper authentication had to be provided to the password-protected device and the SIM card had to be inserted before contents were successfully acquired. Basic subscriber and service provider information was found and reported (i.e., IMEI, IMSI). Memory size is not reported. (Meet)

PIM Applications: All PIM data was found (i.e., Address Book, Calendar, Tasks) in the corresponding database/folder (i.e., Contacts, Calendar folders, Tasks) and reported. Deleted PIM data was not found. (Below)

Dialed/Received Phone Calls: All dialed/received phone calls were found and reported in the Calls folder. Deleted phone calls were not found. (Below)

SMS/MMS Messaging: All active incoming and outgoing SMS messages were found in the SMS folder and reported. Deleted SMS messages were not found. MMS messages and attachments (i.e., graphics, sound bytes) were not found or reported. (Below)

Internet Messaging: Data content associated with sent and received email messages was not found. (Miss)

Web Applications: Visited URLs, search queries performed, textual Web content and graphical images of visited sites were not found. (Miss)

Text File Formats: N.A. – The Sony Ericsson T68i does not support text files (e.g., .txt, .doc, .pdf). (NA)

Graphics Files Format: Supported graphic files (i.e., .jpg, .gif) present on the device were not found. (Miss)

Compressed File Archive Formats: N.A. – The Sony Ericsson T68i does not support compressed archive files (e.g., .zip, .rar, .exe, .tgz). (NA)

Misnamed Files: N.A. – The Sony Ericsson T68i does not support misnamed files (e.g., .txt file renamed with a .dll extension). (NA)

Peripheral Memory Cards: N.A. – The Sony Ericsson T68i does not allow for removable media. (NA)

Acquisition Consistency: N.A. – This signature is not a hash of the device data, but the identity of the examiner. (NA)

Cleared Devices: A Hard Reset was performed by selecting Reset Settings and Reset All in the settings menu. No data was found. (Meet)

Power Loss: The Sony Ericsson T68i was repopulated with the above scenarios, then completely drained of all battery power and reacquired. All data was found as reported above. (Above)

Motorola C333

The following scenarios were conducted on an unlocked GSM Motorola C333. GSM .XRY version 3.0 was used for acquisition.

Connectivity and Retrieval: The password-protected device contents were successfully acquired with or without the SIM present without having to provide proper authentication. Basic subscriber and service provider information was found and reported (i.e., IMEI, IMSI). Memory size is not reported. (Meet)

PIM Applications: All PIM data was found (i.e., Address Book, Calendar) in the corresponding database/folder (i.e., Contacts, Calendar folders) and reported. Deleted PIM data was not found. (Below)

Dialed/Received Phone Calls: All dialed/received phone calls were found and reported in the Calls folder. Deleted phone calls were not found. (Below)

SMS/MMS Messaging: All active incoming and outgoing SMS messages were found in the SMS folder and reported. Deleted SMS messages were not found. MMS messaging is not supported. (Below)

Internet Messaging: N.A. – The Motorola C333 does not support email. (NA)

Web Applications: N.A. – Internet connectivity was unable to be established. (NA)

Text File Formats: N.A. – The Motorola C333 does not support text files (e.g., .txt, .doc, .pdf). (NA)

Graphics Files Format: N.A. – The Motorola C333 does not support graphic files (e.g., .bmp, .jpg, .gif, .png, .tif). (NA)

Compressed File Archive Formats: N.A. – The Motorola C333 does not support compressed archive files (e.g., .zip, .rar, .exe, .tgz). (NA)

Misnamed Files: N.A. – The Motorola C333 does not support misnamed files (e.g., .txt file renamed with a .dll extension). (NA)

Peripheral Memory Cards: N.A. – The Motorola C333 does not allow for removable media. (NA)

Acquisition Consistency: N.A. – This signature is not a hash of the device data, but the identity of the examiner. (NA)

Cleared Devices: A Hard Reset was performed by selecting Master Clear and Master Reset in the settings menu. SMS messages were found and reported in the SMS folder. (Above)

Power Loss: The Motorola C333 was repopulated with the above scenarios, then completely drained of all battery power and reacquired. All data was found as reported above. (Above)

Motorola V6

The following scenarios were conducted on an unlocked GSM Motorola V.series 66. GSM .XRY version 3.0 was used for acquisition.

Connectivity and Retrieval: The password-protected device contents were successfully acquired with or without the SIM present without having to provide proper authentication. Basic subscriber and service provider information was found and reported (i.e., IMEI, IMSI). Memory size is not reported. (Meet)

PIM Applications: All PIM data was found (i.e., Address Book, Calendar) in the corresponding database/folder (i.e., Contacts, Calendar folders) and reported. Deleted PIM data was not found. (Below)

Dialed/Received Phone Calls: All dialed/received phone calls were found and reported in the Calls folder. Deleted phone calls were not found. (Below)

SMS/MMS Messaging: All active incoming and outgoing SMS messages were found in the SMS folder and reported. Deleted SMS messages were not found. MMS messaging is not supported. (Below)

Internet Messaging: N.A. – The Motorola V66 does not support email. (NA)

Web Applications: N.A. – Internet connectivity was unable to be established. (NA)

Text File Formats: N.A. – The Motorola V66 does not support text files (e.g., .txt, .doc, .pdf). (NA)

Graphics Files Format: N.A. – The Motorola V66 does not support graphic files (e.g., .bmp, .jpg, .gif, .png, .tif). (NA)

Compressed File Archive Formats: N.A. – The Motorola V66 does not support compressed archive files (e.g., .zip, .rar, .exe, .tgz). (NA)

Misnamed Files: N.A. – The Motorola V66 does not support misnamed files (e.g., .txt file renamed with a .dll extension). (NA)

Peripheral Memory Cards: N.A. – The Motorola V66 does not allow for removable media. (NA)

Acquisition Consistency: N.A. – This signature is not a hash of the device data, but the identity of the examiner. (NA)

Cleared Devices: A Hard Reset was performed by selecting Master Reset in the settings menu. No data was found. (Meet)

Power Loss: The Motorola V66 was repopulated with the above scenarios, then completely drained of all battery power and reacquired. All data was found as reported above. (Above)

Motorola V300

The following scenarios were conducted on a Pay-As-You-Go GSM Motorola V300. GSM .XRY version 3.0 was used for acquisition.

Connectivity and Retrieval: Proper authentication had to be provided to the password-protected device and the SIM card had to be inserted before contents were successfully acquired. Basic subscriber and service provider information was found and reported (i.e., IMEI, IMSI). Memory size is not reported. (Meet)

PIM Applications: All PIM data was found (i.e., Address Book, Calendar) in the corresponding database/folder (i.e., Contacts, Calendar folders) and reported. Deleted PIM data was not found. (Below)

Dialed/Received Phone Calls: All dialed/received phone calls were found and reported in the Calls folder. Deleted phone calls were not found. (Below)

SMS/MMS Messaging: All active incoming and outgoing SMS messages were found in the SMS folder and reported. Deleted SMS messages were not found. MMS messages and attachments (i.e., graphics, sound bytes) were not found or reported. (Below)

Internet Messaging: Incoming/outgoing emails were found and reported. Chat logs and Deleted messages were not found. (Below)

Web Applications: Visited URLs, search queries performed, textual Web content and graphical images of visited sites were not found. (Miss)

Text File Formats: Data content associated with text files (i.e., .txt, .doc, .pdf) was not found when sent via email. The filename and subject line of the email were found and reported. (Miss)

Graphics Files Format: A connection could not be established allowing the transfer of graphic files (i.e., .bmp, .jpg, .gif .png, .tif) to the Motorola V300. Images were created via the picture camera. No data was found. (Miss)

Compressed File Archive Formats: N.A. – The Motorola V300 does not support compressed archive files (e.g., .zip, .rar, .exe, .tgz). (NA)

Misnamed Files: Misnamed file (e.g., .txt file renamed with a .dll extension) data content was not found when sent via email. The filename and subject line of the email were found and reported. (Miss)

Peripheral Memory Cards: N.A. – The Motorola V300 does not allow for removable media. (NA)

Acquisition Consistency: N.A. – This signature is not a hash of the device data, but the identity of the examiner. (NA)

Cleared Devices: A Hard Reset was performed by selecting the Master Clear and Master Reset options in the settings menu. SMS Messages were recovered. (Above)

Power Loss: The Motorola V300 was repopulated with the above scenarios, then completely drained of all battery power and reacquired. All data was found as reported above. (Above)

Nokia 6610i

The following scenarios were conducted on an unlocked GSM Nokia 6610i. GSM .XRY version 3.0 was used for acquisition.

Connectivity and Retrieval: Proper authentication had to be provided to the password-protected device before contents were successfully acquired. The device contents were successfully acquired with or without the SIM present. Basic subscriber and service provider information was found and reported (i.e., IMEI, IMSI). Memory size is not reported. (Meet)

PIM Applications: All PIM data was found (i.e., Address Book, Calendar) in the corresponding database/folder (i.e., Contacts, Calendar) and reported. Deleted PIM data was not found. (Below)

Dialed/Received Phone Calls: All dialed/received phone calls were found and reported in the Calls folder. Deleted phone calls were not found. (Below)

SMS/MMS Messaging: All active incoming and outgoing SMS messages were found in the SMS Message folder and reported. Deleted SMS/MMS messages were not found. Textual data remnants of received MMS messages were found in the Files folder. Attached MMS data was found in the Picture and Audio folders. (Below)

Internet Messaging: N.A. – The Nokia 6610i does not support email. (NA)

Web Applications: Visited URLs and search engine queries were found and reported. Textual Web content pertaining to URLs was not found. No graphical images of visited sites were found or displayed. (Below)

Text File Formats: Data content associated with text files (i.e., .txt, .doc, .pdf) was found and reported in the Files folder and could be viewed after saved to the forensic workstation. Deleted text files were not found. (Below)

Graphics Files Format: Graphic files (i.e., .bmp, .jpg, .gif, .png, .tif) present on the device were found and reported in the Pictures folder. Deleted graphic files were not found. (Below)

Compressed File Archive Formats: Compressed data file (i.e., .zip, .rar, .exe, .tgz) content was found and reported in the Files folder and could be viewed after saved to the forensic workstation. (Meet)

Misnamed Files: Misnamed files (e.g., .txt file renamed with a .dll extension) were found and reported in the Files folder and could be viewed with the proper application after saved to the forensic workstation. (Meet)

Peripheral Memory Cards: N.A. – The Nokia 6610i does not allow for removable media. (NA)

Acquisition Consistency: N.A. – This signature is not a hash of the device data, but the identity of the examiner. (NA)

Cleared Devices: N.A. – A Hard Reset function is not provided by the phone. (NA)

Power Loss: The Nokia 6610i was repopulated with the above scenarios, then completely drained of all battery power and reacquired. All data was found as reported above. (Above)

Nokia 6200

The following scenarios were conducted on a GSM Nokia 6200. GSM .XRY version 3.0 was used for acquisition.

Connectivity and Retrieval: Proper authentication had to be provided to the password-protected device before contents were successfully acquired. The device contents were successfully acquired with or without the SIM present. Basic subscriber and service provider information was found and reported (i.e., IMEI). Memory size is not reported. (Meet)

PIM Applications: All PIM data was found (i.e., Address Book, Calendar, Tasks) in the corresponding database/folder (i.e., Contacts, Calendar, Notes) and reported. Deleted PIM data was not found. (Below)

Dialed/Received Phone Calls: Dialed/received phone calls were found and reported. Deleted phone calls were not found. (Below)

SMS/MMS Messaging: All active incoming and outgoing SMS text messages were found and reported. Deleted SMS/MMS messages were not found. MMS messages with attachments (i.e., graphics, audio files) were found. (Below)

Internet Messaging: N.A. The Nokia 6200 does not support email. (NA)

Web Applications: Visited URLs, search queries performed, textual Web content and graphical images of visited sites were not found. (Miss)

Text File Formats: Data content associated with text files (i.e., .txt, .doc, .pdf) was found and reported in the Files folder and could be viewed after saved to the forensic workstation. Deleted text files were not found. (Below)

Graphics Files Format: Graphic files (i.e., .bmp, .jpg, .gif, .png, .tif) present on the device were found and reported in the Pictures folder. Deleted graphic files were not found. (Below)

Compressed File Archive Formats: Compressed data file (i.e., .zip, .rar, .exe, .tgz) content was found and reported in the Files folder and could be viewed after saved to the forensic workstation. (Meet)

Misnamed Files: Misnamed files (e.g., .txt file renamed with a .dll extension) were found and reported in the Files folder and could be viewed with the proper application after saved to the forensic workstation. (Meet)

Peripheral Memory Cards: N.A. – The Nokia 6200 does not allow for removable media. (NA)

Acquisition Consistency: N.A. – This signature is not a hash of the device data, but the identity of the examiner. (NA)

Cleared Devices: N.A. – A Hard Reset function is not provided by the phone. (NA)

Power Loss: The Nokia 6200 was repopulated with the above scenarios, then completely drained of all battery power and reacquired. All data was found as reported above. (Above)

Nokia 7610

The following scenarios were conducted on a GSM Nokia 7610 running Symbian OS. GSM .XRY version 3.0 was used for acquisition.

Connectivity and Retrieval: GSM .XRY failed to acquire data contents via a cable interface. Therefore, Bluetooth was used for acquisition. Proper authentication had to be provided to the password-protected device in order to turn Bluetooth on to allow for connectivity with the GSM .XRY unit. The SIM must be present for acquisition, but authentication did not have to be provided. Basic subscriber and service provider information was found and reported (i.e., IMEI, IMSI). Memory size is not reported. (Meet)

PIM Applications: All PIM data was found (i.e., Address Book, Calendar, Tasks) in the corresponding database/folder (i.e., Contacts, Calendar, Notes) and reported. Deleted PIM data was not found. (Below)

Dialed/Received Phone Calls: Dialed/received phone calls were not found. (Miss)

SMS/MMS Messaging: Incoming and outgoing SMS/MMS text messages (i.e., text based content) were not found. (Miss)

Internet Messaging: Data content associated with sent and received email messages was not found. (Miss)

Web Applications: Visited URLs, search queries performed, textual Web content and graphical images of visited sites were not found. (Miss)

Text File Formats: Data content associated with text files (i.e., .txt. pdf, .doc) were found and reported in the Notes folder. Deleted text file were not found. (Below)

Graphics Files Format: Graphic files (e.g., .bmp, .jpg, .gif, .png, .tif) present on the device were found and reported in the Pictures folder. Deleted graphic files were not found. (Below)

Compressed File Archive Formats: Compressed data file (i.e., .zip, .rar, .exe, .tgz) content was found and reported in the Files folder and could be viewed after saved to the forensic workstation. (Meet)

Misnamed Files: Misnamed files (e.g., .txt file renamed with a .dll extension) were found and reported in the Notes folder. (Meet)

Peripheral Memory Cards: Data residing on a 64 MB MMC populated with various files (i.e., text, graphics, audio, misnamed files) were found and reported. Deleted data was not found. (Below)

Acquisition Consistency: N.A. – This signature is not a hash of the device data, but the identity of the examiner. (NA)

Cleared Devices: A Hard Reset was performed by entering *#7370# followed by the call key. No internal phone memory data was found. Data populated onto the MMC card was found and reported. (Meet)

Power Loss: The Nokia 7610 was repopulated with the above scenarios, then completely drained of all battery power and reacquired. All data was found as reported above. (Above)

Appendix D: SecureView Results

The scenarios were performed on a forensic workstation running Windows XP SP2. SecureView version 1.5.0 was used to acquire data from the following cell phones: Audiovox 8910, Ericsson T68i, LG4015, Motorola C333, Motorola V66, Motorola V300, Nokia 6200, and a Sanyo 8200 via a data-link cable.

Audiovox 8910

The following scenarios were conducted on a pre-paid CMDA Audiovox 8910.

Connectivity and Retrieval: The password-protected device contents were successfully acquired when authentication mechanisms were provided with the SIM present. No data was found. (Miss)

Ericsson T68i

The following scenarios were conducted on an unlocked Sony Ericsson T68i. Connectivity was established using Susteen's Sony Ericsson Rabbit cable.

Connectivity and Retrieval: The password-protected device contents were successfully acquired when authentication mechanisms were provided with the SIM present. Basic subscriber information (i.e. IMEI/ESN) was not found. Memory size is not reported. (Meet)

PIM Applications: All active phone book entries stored in internal phone memory were found and reported. Entries stored on the SIM were not reported. Active calendar entries were found. Deleted PIM data was not found. (Below)

Dialed/Received Phone Calls: Dialed/received phone calls were not found. (Miss)

SMS/MMS Messaging: All active incoming and outgoing SMS messages were found and reported. Deleted SMS messages were not found. MMS messages and attachments (i.e., graphics, sound bytes) were not found or reported. (Below)

Internet Messaging: Data content associated with sent and received email messages was not found. (Miss)

Web Applications: Visited URLs, search queries performed, textual Web content and graphical images of visited sites were not found. (Miss)

Text File Formats: N.A. – The Sony Ericsson T68i does not support text files (e.g., .txt, .doc, .pdf). (NA)

Graphics Files Format: Supported graphic files (i.e., .jpg, .gif) present on the device were not found. (Miss)

Compressed File Archive Formats: N.A. – The Sony Ericsson T68i does not support compressed archive files (e.g., .zip, .rar, .exe, .tgz). (NA)

Misnamed Files: N.A. – The Sony Ericsson T68i does not support misnamed files (e.g., .txt file renamed with a .dll extension). (NA)

Peripheral Memory Cards: N.A. – The Sony Ericsson T68i does not allow for removable media. (NA)

Acquisition Consistency: N.A. – SecureView does not provide an internal hashing algorithm for individual files or overall phone acquisition. (NA)

Cleared Devices: A Hard Reset was performed by selecting Master Reset in the settings menu. Data contained on the SIM was found and reported. (Meet)

Power Loss: The Sony Ericsson T68i was repopulated with the above scenarios, then completely drained of all battery power and reacquired. All data was found as reported above. (Above)

LG4015

The following scenarios were conducted on a LG4015. Connectivity was established using Susteen's LG2 Camel cable.

Connectivity and Retrieval: The password-protected device contents were successfully acquired. Basic subscriber information (i.e. IMEI/ESN) was found. Memory size is not reported. (Meet)

PIM Applications: All active phone book entries stored in internal phone memory and the SIM were found and reported. Active calendar entries were found. Deleted PIM data was not found. (Below)

Dialed/Received Phone Calls: Dialed/received phone calls were not found. (Miss)

SMS/MMS Messaging: No data was found. (Miss)

Internet Messaging: N.A. – The LG 4015 does not support email. (NA)

Web Applications: No data was found. (Miss)

Text File Formats: N.A. – The LG 4015 does not support text files (e.g., .txt, .doc, .pdf). (NA)

Graphics Files Format: N.A. – A connection could not be established allowing the transfer of graphic files (i.e., .bmp, .jpg, .gif .png, .tif) to the LG 4015. (NA)

Compressed File Archive Formats: N.A. – The LG 4015 does not support compressed archive files (e.g., .zip, .rar, .exe, .tgz). (NA)

Misnamed Files: N.A. – The LG 4015 does not support misnamed files (e.g., .txt file renamed with a .dll extension). (NA)

Peripheral Memory Cards: N.A. – The LG 4015 does not allow for removable media. (NA)

Acquisition Consistency: N.A. – SecureView does not provide an internal hashing algorithm for individual files or overall phone acquisition. (NA)

Cleared Devices: A Hard Reset was performed by selecting the Reset option in the security menu. All active data was recovered. (Above)

Power Loss: The Sanyo PM-8200 was repopulated with the above scenarios, then completely drained of all battery power and reacquired. All data was found as reported above. (Above)

Motorola C333

The following scenarios were conducted on an unlocked Motorola C333 GSM phone. Connectivity was established using Susteen's Motorola 2 Scorpion cable.

Connectivity and Retrieval: The password-protected device contents were successfully acquired with SIM present and providing proper authentication. Basic subscriber information (i.e. IMEI/ESN) was found and reported. Memory size is not reported. (Meet)

PIM Applications: All active phone book entries were found and reported. Active calendar entries were found. Deleted PIM data was not found. (Below)

Dialed/Received Phone Calls: Dialed/received phone calls were not found. (Miss)

SMS/MMS Messaging: All active incoming and outgoing SMS messages were found and reported. Deleted SMS messages were not found. MMS messages and attachments (i.e., graphics, sound bytes) were not found or reported. (Below)

Internet Messaging: N.A. – The Motorola C333 does not support email. (NA)

Web Applications: N.A. – Internet connectivity was unable to be established. (NA)

Text File Formats: N.A. – The Motorola C333 does not support text files (e.g., .txt, .doc, .pdf). (NA)

Graphics Files Format: N.A. – The Motorola C333 does not support graphic files (e.g., .bmp, .jpg, .gif, .png, .tif). (NA)

Compressed File Archive Formats: N.A. – The Motorola C333 does not support compressed archive files (e.g., .zip, .rar, .exe, .tgz). (NA)

Misnamed Files: N.A. – The Motorola C333 does not support misnamed files (e.g., .txt file renamed with a .dll extension). (NA)

Peripheral Memory Cards: N.A. – The Motorola C333 does not allow for removable media. (NA)

Acquisition Consistency: N.A. – SecureView does not provide an internal hashing algorithm for individual files or overall phone acquisition. (NA)

Cleared Devices: A Hard Reset was performed by selecting the Master Reset option in the settings menu. No data was found. (Meet)

Power Loss: The Motorola C333 was repopulated with the above scenarios, then completely drained of all battery power and reacquired. All data was found as reported above. (Above)

Motorola V6

The following scenarios were conducted on an unlocked Motorola V.series 66 GSM phone. Connectivity was established using Susteen's Motorola 2 Penguin cable.

Connectivity and Retrieval: The password-protected device contents were successfully acquired with the SIM present. Basic subscriber information (i.e. IMEI/ESN) was found and reported. Memory size is not reported. (Meet)

PIM Applications: All active phone book entries were found and reported. Active calendar entries were found. Deleted PIM data was not found. (Below)

Dialed/Received Phone Calls: Dialed/received phone calls were not found. (Miss)

SMS/MMS Messaging: All active incoming and outgoing SMS messages were found and reported. Deleted SMS messages were not found. MMS messaging is not supported. (Below)

Internet Messaging: N.A. – The Motorola V66 does not support email. (NA)

Web Applications: N.A. – Internet connectivity was unable to be established. (NA)

Text File Formats: N.A. – The Motorola V66 does not support text files (e.g., .txt, .doc, .pdf). (NA)

Graphics Files Format: N.A. – The Motorola V66 does not support graphic files (e.g., .bmp, .jpg, .gif, .png, .tif). (NA)

Compressed File Archive Formats: N.A. – The Motorola V66 does not support compressed archive files (e.g., .zip, .rar, .exe, .tgz). (NA)

Misnamed Files: N.A. – The Motorola V66 does not support misnamed files (e.g., .txt file renamed with a .dll extension). (NA)

Peripheral Memory Cards: N.A. – The Motorola V66 does not allow for removable media. (NA)

Acquisition Consistency: N.A. – SecureView does not provide an internal hashing algorithm for individual files or overall phone acquisition. (NA)

Cleared Devices: A Hard Reset was performed by selecting the Master Clear option in the settings menu. No data was found. (Meet)

Power Loss: The Motorola V66 was repopulated with the above scenarios, then completely drained of all battery power and reacquired. All data was found as reported above. (Above)

Motorola V300

The following scenarios were conducted on an unlocked Motorola V300 GSM phone. Connectivity was established using Susteen's Motorola 2 Penguin cable.

Connectivity and Retrieval: Proper authentication had to be provided to the password-protected device and the SIM card had to be inserted before contents were successfully acquired. Basic subscriber information (i.e. IMEI/ESN) was found and reported. Memory size is not reported. (Meet)

PIM Applications: All active phone book entries were found and reported. Active calendar entries were found. Deleted PIM data was not found. (Below)

Dialed/Received Phone Calls: Dialed/received phone calls were not found. (Miss)

SMS/MMS Messaging: All active incoming and outgoing SMS messages were found and reported. Deleted SMS and MMS message data (i.e., images, sound bytes) were not found. (Below)

Internet Messaging: No data was found. (Miss)

Web Applications: Visited URLs, search queries performed, textual Web content and graphical images of visited sites were not found. (Miss)

Text File Formats: Data content associated with text files (e.g., .txt, .doc, .pdf) was not found. (Miss)

Graphics Files Format: A connection could not be established allowing the transfer of graphic files (e.g., .bmp, .jpg, .gif .png, .tif) to the Motorola V300. Images were created by using the picture camera. Active data was found and reported. (Meet).

Compressed File Archive Formats: N.A. – The Motorola V300 does not support compressed archive files (e.g., .zip, .rar, .exe, .tgz). (NA)

Misnamed Files: N.A. – The Motorola V300 does not support misnamed files (e.g., .txt file renamed with a .dll extension). (NA)

Peripheral Memory Cards: N.A. – The Motorola V300 does not allow for removable media. (NA)

Acquisition Consistency: N.A. – SecureView does not provide an internal hashing algorithm for individual files or overall phone acquisition. (NA)

Cleared Devices: A Hard Reset was performed by selecting the Master Reset option in the settings menu. No data was found. (Meet)

Power Loss: The Motorola V300 was repopulated with the above scenarios, then completely drained of all battery power and reacquired. All data was found as reported above. (Above)

Nokia 6200

The following scenarios were conducted on a Nokia 6200 GSM phone. Connectivity was established using Susteen's Nokia 2 Cobra cable.

Connectivity and Retrieval: The password-protected device contents were successfully acquired with SIM present and providing proper authentication. Basic subscriber information (i.e. IMEI/ESN) was found and reported. Memory size is not reported. (Meet)

PIM Applications: All active phone book entries were found and reported. Active calendar entries were found. Deleted PIM data was not found. (Below)

Dialed/Received Phone Calls: Dialed/received phone calls were not found. (Miss)

SMS/MMS Messaging: All active incoming and outgoing SMS messages were found and reported. Active MMS and deleted SMS/MMS messages were not found. (Below)

Internet Messaging: N.A. – The Nokia 6200 does not support email. (NA)

Web Applications: Visited URLs, search queries performed, textual Web content and graphical images of visited sites were not found. (Miss)

Text File Formats: Data content associated with text files (i.e., .txt, .doc, .pdf) was not found. (Miss)

Graphics Files Format: Graphic files (i.e., .bmp, .jpg, .gif, .png, .tif) present on the device were found and reported in the Pictures folder. Deleted graphic files were not found. (Below)

Compressed File Archive Formats: Compressed data file (i.e., .zip, .rar, .exe, .tgz) content was not found. (Miss)

Misnamed Files: Misnamed files (e.g., `.txt` file renamed with a `.dll` extension) were not found. (Miss)

Peripheral Memory Cards: N.A. – The Nokia 6200 does not allow for removable media. (NA)

Acquisition Consistency: N.A. – SecureView does not provide an internal hashing algorithm for individual files or overall phone acquisition. (NA)

Cleared Devices: N.A. – A Hard Reset function is not provided by the phone. (NA)

Power Loss: The Nokia 6200 was repopulated with the above scenarios, then completely drained of all battery power and reacquired. All data was found as reported above. (Above)

Sanyo PM-8200

The following scenarios were conducted on a CMDA Sanyo 8200. Connectivity was established using Susteen's Sanyo "Dog" cable.

Connectivity and Retrieval: The password-protected device contents were successfully acquired. Basic subscriber information (i.e. IMEI/ESN) was found and reported. Memory size is not reported. (Meet)

PIM Applications: All active phone book entries were found and reported. Active calendar entries were found. Deleted PIM data was not found. (Below)

Dialed/Received Phone Calls: Dialed/received phone calls were not found. (Miss)

SMS/MMS Messaging: SMS/MMS messages were not found. (Miss)

Internet Messaging: Data content associated with sent and received email messages was not found. (Miss)

Web Applications: Data content associated with web applications (i.e., visited URLs, graphics, etc.) were not found. (Miss)

Text File Formats: N.A. – The Sanyo PM-8200 does not support text files (e.g., `.txt`, `.doc`, `.pdf`). (NA)

Graphics Files Format: A connection could not be established allowing the transfer of graphic files (i.e., `.bmp`, `.jpg`, `.gif` `.png`, `.tif`) to the Sanyo PM-8200. Images were created via the picture camera. Graphic files present on the device were found and reported. Deleted graphic files were not found. (Below)

Compressed File Archive Formats: N.A. – The Sanyo PM-8200 does not support compressed archive files (e.g., `.zip`, `.rar`, `.exe`, `.tgz`). (NA)

Misnamed Files: N.A. – The Sanyo PM-8200 does not support misnamed files (e.g., .txt file renamed with a .dll extension). (NA)

Peripheral Memory Cards: N.A. – The Sanyo PM-8200 does not allow for removable media. (NA)

Acquisition Consistency: N.A. – SecureView does not provide an internal hashing algorithm for individual files or overall phone acquisition. (NA)

Cleared Devices: A Hard Reset was performed by selecting the Reset option in the security menu. All active data is recovered. (Above)

Power Loss: The Sanyo PM-8200 was repopulated with the above scenarios, then completely drained of all battery power and reacquired. All data was found as reported above. (Above)

Appendix E: PhoneBase2 Results

The scenarios were performed on a forensic workstation running Windows XP SP2. Phonebase2 version 1.2.0.15 was used to acquire data from the following cell phones: Ericsson T68i, Motorola V66, Motorola V300 and a Nokia 6610i via a data-link cable.

Ericsson T68i

The following scenarios were conducted on a Sony Ericsson T68i.

Connectivity and Retrieval: Proper authentication had to be provided to the password-protected device and the SIM card had to be inserted before contents were successfully acquired. Basic subscriber and service provider information was found and reported (i.e., IMEI). Memory size is not reported. (Meet)

PIM Applications: Partial PIM data was found and reported (i.e., Address Book). Calendar, Task entries and deleted PIM data were not found. (Below)

Dialed/Received Phone Calls: All dialed/received phone calls were found and reported. Deleted phone calls were not found. (Below)

SMS/MMS Messaging: All active incoming and outgoing SMS messages were found and reported. Deleted SMS messages were not found. MMS messages and attachments (i.e., graphics, sound bytes) were not found or reported. (Below)

Internet Messaging: Data content associated with sent and received email messages were not found. (Miss)

Web Applications: Visited URLs, search queries performed, textual Web content and graphical images of visited sites were not found. (Miss)

Text File Formats: N.A. – The Sony Ericsson T68i does not support text files (e.g., .txt, .doc, .pdf). (NA)

Graphics Files Format: Supported graphic files (i.e., .jpg, .gif) present on the device were not found. (Miss)

Compressed File Archive Formats: N.A. – The Sony Ericsson T68i does not support compressed archive files (e.g., .zip, .rar, .exe, .tgz). (NA)

Misnamed Files: N.A. – The Sony Ericsson T68i does not support misnamed files (e.g., .txt file renamed with a .dll extension). (NA)

Peripheral Memory Cards: N.A. – The Sony Ericsson T68i does not allow for removable media. (NA)

Acquisition Consistency: N.A. – PhoneBase2 does not provide an internal hashing algorithm for individual files or overall phone acquisition. (NA)

Cleared Devices: A Hard Reset was performed by selecting Master Reset in the settings menu. No data was found. (Meet)

Power Loss: The Sony Ericsson T68i was repopulated with the above scenarios, then completely drained of all battery power and reacquired. All data was found as reported above. (Above)

Motorola V6

The following scenarios were conducted on an unlocked Motorola V.series 66 GSM phone. Connectivity was established using Susteen's Motorola 2 Penguin cable.

Connectivity and Retrieval: The password-protected device contents were successfully acquired with or without the SIM present. Basic subscriber information (i.e. IMEI) was found and reported. Memory size is not reported. (Meet)

PIM Applications: Partial PIM data was found and reported (i.e., Address Book). Calendar, Task entries and deleted PIM data was not found (Below)

Dialed/Received Phone Calls: All dialed/received phone calls were found and reported. Deleted phone calls were not found. (Below)

SMS/MMS Messaging: All active incoming and outgoing SMS messages were found and reported. Deleted SMS messages were not found. MMS messaging is not supported. (Below)

Internet Messaging: N.A. – The Motorola V66 does not support email. (NA)

Web Applications: N.A. – Internet connectivity was unable to be established. (NA)

Text File Formats: N.A. – The Motorola V66 does not support text files (e.g., .txt, .doc, .pdf). (NA)

Graphics Files Format: N.A. – The Motorola V66 does not support graphic files (e.g., .bmp, .jpg, .gif, .png, .tif). (NA)

Compressed File Archive Formats: N.A. – The Motorola V66 does not support compressed archive files (e.g., .zip, .rar, .exe, .tgz). (NA)

Misnamed Files: N.A. – The Motorola V66 does not support misnamed files (e.g., .txt file renamed with a .dll extension). (NA)

Peripheral Memory Cards: N.A. – The Motorola V66 does not allow for removable media. (NA)

Acquisition Consistency: N.A. – PhoneBase2 does not provide an internal hashing algorithm for individual files or overall phone acquisition. (NA)

Cleared Devices: A Hard Reset was performed by selecting the Master Clear option in the settings menu. No data was found. (Meet)

Power Loss: The Motorola V66 was repopulated with the above scenarios, then completely drained of all battery power and reacquired. All data was found as reported above. (Above)

Motorola V300

The following scenarios were conducted on an unlocked Motorola V300 GSM phone. Connectivity was established using Susteen's Motorola 2 Penguin cable.

Connectivity and Retrieval: Proper authentication had to be provided to the password-protected device and the SIM card had to be inserted before contents were successfully acquired. Basic subscriber information (i.e. IMEI) was found and reported. Memory size is not reported. (Meet)

PIM Applications: Partial PIM data was found and reported (i.e., Address Book). Calendar, Task entries and deleted PIM data were not found (Below)

Dialed/Received Phone Calls: All dialed/received phone calls were found and reported. Deleted phone calls were not found. (Below)

SMS/MMS Messaging: All active incoming and outgoing SMS messages were found and reported. Deleted SMS and MMS message data (i.e., images, sound bytes) were not found. (Below)

Internet Messaging: No data was found. (Miss)

Web Applications: Visited URLs, search queries performed, textual Web content and graphical images of visited sites were not found. (Miss)

Text File Formats: Data content associated with text files (e.g., .txt, .doc, .pdf) was not found. (Miss)

Graphics Files Format: A connection could not be established allowing the transfer of graphic files (e.g., .bmp, .jpg, .gif .png, .tif) to the Motorola V300. Images were created by using the picture camera. Active data was found and reported. (Meet).

Compressed File Archive Formats: N.A. – The Motorola V300 does not support compressed archive files (e.g., .zip, .rar, .exe, .tgz). (NA)

Misnamed Files: N.A. – The Motorola V300 does not support misnamed files (e.g., .txt file renamed with a .dll extension). (NA)

Peripheral Memory Cards: N.A. – The Motorola V300 does not allow for removable media. (NA)

Acquisition Consistency: N.A. – PhoneBase2 does not provide an internal hashing algorithm for individual files or overall phone acquisition. (NA)

Cleared Devices: A Hard Reset was performed by selecting the Master Reset option in the settings menu. No data was found. (Meet)

Power Loss: The Motorola V300 was repopulated with the above scenarios, then completely drained of all battery power and reacquired. All data was found as reported above. (Above)

Nokia 6610i

The following scenarios were conducted on a Nokia 6610i via IrDA.

Connectivity and Retrieval: Proper authentication had to be provided to the password-protected device and the SIM card had to be inserted before contents were successfully acquired via IrDA. PhoneBase2 does not support a data link cable interface for the Nokia 6610i. Basic subscriber and service provider information was found and reported (i.e., IMEI). Memory size is not reported. (Meet)

PIM Applications: Partial PIM data was found and reported (i.e., Address Book). Calendar entries, Tasks and deleted PIM data were not found. (Below)

Dialed/Received Phone Calls: All dialed/received phone calls were found and reported. Deleted phone calls were not found. (Below)

SMS/MMS Messaging: All active incoming and outgoing SMS messages were found and reported. Active MMS and deleted SMS/MMS messages were not found. (Below)

Internet Messaging: N.A. – The Nokia 6610i does not support email. (NA)

Web Applications: Visited URLs, search queries performed, textual Web content and graphical images of visited sites were not found. (Miss)

Text File Formats: Data content associated with text files (i.e., .txt, .doc, .pdf) was found and reported with an appropriate third party application. (Meet)

Graphics Files Format: Graphic files (e.g., .bmp, .jpg, .gif, .png, .tif) were found and reported with an appropriate third party application. Deleted graphic file data was not found. (Below)

Compressed File Archive Formats: Compressed data file (i.e., .zip, .rar, .exe, .tgz) content was found and reported with an appropriate third party application. (Meet)

Misnamed Files: Misnamed files (e.g., `.txt` file renamed with a `.dll` extension) were found and reported with an appropriate third party application. (Meet)

Peripheral Memory Cards: N.A. – The Nokia 6610i does not allow for removable media. (NA)

Acquisition Consistency: N.A. – PhoneBase2 does not provide an internal hashing algorithm for individual files or overall phone acquisition. (NA)

Cleared Devices: N.A. – A Hard Reset function is not provided by the phone. (NA)

Power Loss: The Nokia 6610i was repopulated with the above scenarios, then completely drained of all battery power and reacquired. All data was found as reported above. (Above)

Appendix F: CellDEK Results

The scenarios were performed on the CellDEK forensic workstation. CellDEK versions 1.3.0.0 – 1.3.1.0 were used to acquire data from the following cell phones: Ericsson T68i, Motorola MPX220, Motorola V66, Motorola V300, Nokia 6610i and a Nokia 6200.

Ericsson T68i

The following scenarios were conducted on a Sony Ericsson T68i via a data link cable. CellDEK version 1.3.0.0 was used for acquisition.

Connectivity and Retrieval: Proper authentication had to be provided to the password-protected device and the SIM card had to be inserted before contents were successfully acquired. Basic subscriber and service provider information was found and reported (i.e., IMEI). Memory size is not reported. (Meet)

PIM Applications: All PIM data was found and reported (i.e., Address Book, Calendar, Tasks). Deleted PIM data was not found. (Below)

Dialed/Received Phone Calls: All dialed/received phone calls were found and reported. Deleted phone calls were not found. (Below)

SMS/MMS Messaging: All active incoming and outgoing SMS messages were found and reported. Deleted SMS messages were not found. MMS messages and attachments (i.e., graphics, sound bytes) were not found or reported. (Below)

Internet Messaging: Data content associated with sent and received email messages were not found. (Miss)

Web Applications: Visited URLs, search queries performed, textual Web content and graphical images of visited sites were not found. (Miss)

Text File Formats: N.A. – The Sony Ericsson T68i does not support text files (e.g., .txt, .doc, .pdf). (NA)

Graphics Files Format: Supported graphic files (i.e., .jpg, .gif) present on the device were not found. (Miss)

Compressed File Archive Formats: N.A. – The Sony Ericsson T68i does not support compressed archive files (e.g., .zip, .rar, .exe, .tgz). (NA)

Misnamed Files: N.A. – The Sony Ericsson T68i does not support misnamed files (e.g., .txt file renamed with a .dll extension). (NA)

Peripheral Memory Cards: N.A. – The Sony Ericsson T68i does not allow for removable media. (NA)

Acquisition Consistency: CellDEK does not provide an overall hash for the case file. However, hashes for individual data elements are consistent after back to back acquisitions. (Meet)

Cleared Devices: A Hard Reset was performed by selecting Master Reset in the settings menu. Data contained on the SIM was found and reported. (Meet)

Power Loss: The Sony Ericsson T68i was repopulated with the above scenarios, then completely drained of all battery power and reacquired. All data was found as reported above. (Above)

Motorola MPx2

The following scenarios were conducted on a Cingular GSM Motorola MPx220 running Microsoft Windows Mobile 2004 for Pocket PC Phone Edition. CellDEK version 1.3.1.0 was used for acquisition.

Connectivity and Retrieval: Proper authentication had to be provided to the password-protected device before contents were successfully acquired. Basic subscriber and service provider information was found and reported (i.e., IMSI, IMEI) with or without the SIM present. Memory size is not reported. (Meet)

PIM Applications: Partial PIM data was found and reported (i.e., Address Book, Calendar). Deleted PIM data and Tasks were not found. (Below)

Dialed/Received Phone Calls: All dialed/received phone calls were found and reported. Deleted phone calls were not found. (Below)

SMS/MMS Messaging: All active incoming and outgoing SMS messages were found and reported. Active MMS and deleted SMS/MMS messages were not found. (Below)

Internet Messaging: Active incoming and outgoing email data was found and reported. Deleted data or attachments were not found. (Below)

Web Applications: Visited URLs, search queries performed, textual Web content and graphical images of visited sites were not found. (Miss)

Text File Formats: Data content associated with text files (i.e., .txt, .doc, .pdf) was not found. (Miss)

Graphics File Formats: Graphics file (i.e., .bmp, .jpg, .gif, .png, .tif) data content was found and reported. Deleted graphics files were not found. (Below)

Compressed File Archive Formats: Compressed data file (i.e., .zip, .rar, .exe, .tgz) content was not found. (Miss)

Misnamed Files: Misnamed file (e.g., .txt file renamed with a .dll extension) data content was not found and reported. (Miss)

Peripheral Memory Cards: Data residing on a 256 MB Mini SD Card populated with various files (i.e., text, graphics, audio, compressed archive files, misnamed files) was found and reported. Deleted files were not found. (Below)

Acquisition Consistency: CellDEK does not provide an overall hash for the case file. However, hashes for individual data elements are consistent after back to back acquisitions. (Meet)

Cleared Devices: A Hard Reset was performed by selecting the Master Clear option in the settings menu. No data was found. (Meet)

Power Loss: The Motorola MPx220 was repopulated with the above scenarios, then completely drained of all battery power and reacquired. The following data was found: MMS messages (text content) and Internet Messaging data (i.e., sent/received email). Individual files stored in the /storage directory were found and reported. (Above)

Motorola v6

The following scenarios were conducted on a Motorola v66 via a data link cable. CellDEK version 1.3.1.0 was used for acquisition.

Connectivity and Retrieval: Proper authentication had to be provided when authentication mechanisms were enabled on the SIM. However, data was captured when authentication mechanisms (i.e., internal memory locks) were employed with the SIM present. Acquisition was unsuccessful if the SIM card was absent. Basic subscriber and service provider information was found and reported (i.e., IMSI, IMEI). Memory size is not reported. (Meet)

PIM Applications: Partial PIM data was found and reported (i.e., Address Book, Calendar). Deleted PIM data and Tasks were not found. (Below)

Dialed/Received Phone Calls: All dialed/received phone calls were found and reported. Deleted phone calls were not found. (Below)

SMS/MMS Messaging: All active incoming and outgoing SMS messages were found and reported. Active MMS and deleted SMS/MMS messages were not found. (Below)

Internet Messaging: N.A. – The Motorola v66 does not support email. (NA)

Web Applications: N.A. – Internet connectivity was unable to be established. (NA)

Text File Formats: N.A. – The Motorola V66 does not support text files (e.g., .txt, .doc, .pdf). (NA)

Graphics Files Format: N.A. – The Motorola V66 does not support graphic files (e.g., .bmp, .jpg, .gif, .png, .tif). (NA)

Compressed File Archive Formats: N.A. – The Motorola V66 does not support compressed archive files (e.g., .zip, .rar, .exe, .tgz). (NA)

Misnamed Files: N.A. – The Motorola V66 does not support misnamed files (e.g., .txt file renamed with a .dll extension). (NA)

Peripheral Memory Cards: N.A. – The Motorola v66 does not allow for removable media. (NA)

Acquisition Consistency: CellDEK does not provide an overall hash for the case file. However, hashes for individual data elements are consistent after back to back acquisitions. (Meet)

Cleared Devices: A Hard Reset was performed by selecting the Master Clear option in the settings menu. No data was found. (Meet)

Power Loss: The Motorola v66 was repopulated with the above scenarios, then completely drained of all battery power and reacquired. All data was found as reported above. (Above)

Motorola v30

The following scenarios were conducted on a Motorola v300 via a data link cable. CellDEK version 1.3.1.0 was used for acquisition.

Connectivity and Retrieval: Proper authentication had to be provided to the password-protected device and the SIM card had to be inserted before contents were successfully acquired. However, partial data (i.e., images captured via internal camera) were captured when authentication mechanisms (i.e., internal memory locks, SIM locks) were employed, although Error #1 AT-CCLK was reported on the Data View screen. Basic subscriber and service provider information was found and reported (i.e., IMSI, IMEI). Memory size is not reported. (Meet)

PIM Applications: Partial PIM data was found and reported (i.e., Address Book, Calendar). Deleted PIM data and Tasks were not found. (Below)

Dialed/Received Phone Calls: All dialed/received phone calls were found and reported. Deleted phone calls were not found. (Below)

SMS/MMS Messaging: All active incoming and outgoing SMS messages were found and reported. Active MMS and deleted SMS/MMS messages were not found. (Below)

Internet Messaging: Active incoming and outgoing email data was found and reported. Deleted data or attachments were not found. (Below)

Web Applications: Visited URLs, search queries performed, textual Web content and graphical images of visited sites were not found. (Miss)

Text File Formats: Data content associated with text files (e.g., .txt, .doc, .pdf) was not found. (Miss)

Graphics Files Format: A connection could not be established allowing the transfer of graphic files (e.g., .bmp, .jpg, .gif .png, .tif) to the Motorola V300. Images were created by using the picture camera. Active data was found and reported. (Meet).

Compressed File Archive Formats: N.A. – The Motorola V300 does not support compressed archive files (e.g., .zip, .rar, .exe, .tgz). (NA)

Misnamed Files: N.A. – The Motorola V300 does not support misnamed files (e.g., .txt file renamed with a .dll extension). (NA)

Peripheral Memory Cards: N.A. – The Motorola v66 does not allow for removable media. (NA)

Acquisition Consistency: CellDEK does not provide an overall hash for the case file. However, hashes for individual data elements are consistent after back to back acquisitions. (Meet)

Cleared Devices: A Hard Reset was performed by selecting the Master Clear option in the settings menu. No data was found. (Meet)

Power Loss: The Motorola v300 was repopulated with the above scenarios, then completely drained of all battery power and reacquired. All data was found as reported above. (Above)

Nokia 6200

The following scenarios were conducted on a Nokia 6200 via IrDA. CellDEK version 1.3.1.0 was used for acquisition.

Connectivity and Retrieval: Proper authentication had to be provided to the password-protected device and the SIM card had to be inserted before contents were successfully acquired via IrDA, CellDEK does not provide a data link cable interface for the Nokia 6200. Basic subscriber and service provider information was found and reported (i.e., IMEI). Memory size is not reported. (Meet)

PIM Applications: Partial PIM data was found and reported (i.e., Address Book, Calendar). Deleted PIM data and Tasks were not found. (Below)

Dialed/Received Phone Calls: All dialed/received phone calls were found and reported. Deleted phone calls were not found. (Below)

SMS/MMS Messaging: All active incoming and outgoing SMS messages were found and reported. Active MMS and deleted SMS/MMS messages were not found. (Below)

Internet Messaging: N.A. – The Nokia 6200 does not support email. (NA)

Web Applications: Visited URLs, search queries performed, textual Web content and graphical images of visited sites were not found. (Miss)

Text File Formats: Data content associated with text files (i.e., .txt, .doc, .pdf) was not found. (Miss)

Graphics Files Format: Graphic files (e.g., .bmp, .jpg, .gif, .png) were found and reported. Deleted graphic file data and .tif files were not found. (Below)

Compressed File Archive Formats: Compressed data file (i.e., .zip, .rar, .exe, .tgz) content was not found. (Miss)

Misnamed Files: Misnamed files (e.g., .txt file renamed with a .dll extension) were not found. (Miss)

Peripheral Memory Cards: N.A. – The Nokia 6200 does not allow for removable media. (NA)

Acquisition Consistency: CellDEK does not provide an overall hash for the case file. However, hashes for individual data elements are consistent after back to back acquisitions. (Meet)

Cleared Devices: N.A. – A Hard Reset function is not provided by the phone. (NA)

Power Loss: The Nokia 6200 was repopulated with the above scenarios, then completely drained of all battery power and reacquired. All data was found as reported above. (Above)

Nokia 6610i

The following scenarios were conducted on a Nokia 6610i via IrDA. CellDEK version 1.3.0.0 was used for acquisition.

Connectivity and Retrieval: Proper authentication had to be provided to the password-protected device and the SIM card had to be inserted before contents were successfully acquired via IrDA, CellDEK does not provide a data link cable interface for the Nokia 6610i. Basic subscriber and service provider information was found and reported (i.e., IMEI, IMSI). Memory size is not reported. (Meet)

PIM Applications: Partial PIM data was found and reported (i.e., Address Book, Calendar). Deleted PIM data and Tasks were not found. (Below)

Dialed/Received Phone Calls: All dialed/received phone calls were found and reported. Deleted phone calls were not found. (Below)

SMS/MMS Messaging: All active incoming and outgoing SMS messages were found and reported. Active MMS and deleted SMS/MMS messages were not found. (Below)

Internet Messaging: N.A. – The Nokia 6610i does not support email. (NA)

Web Applications: Visited URLs, search queries performed, textual Web content and graphical images of visited sites were not found. (Miss)

Text File Formats: Data content associated with text files (i.e., .txt, .doc, .pdf) was not found. (Miss)

Graphics Files Format: Graphic files (e.g., .bmp, .jpg, .gif, .png) were found and reported. Deleted graphic file data and .tif files were not found. (Below)

Compressed File Archive Formats: Compressed data file (i.e., .zip, .rar, .exe, .tgz) content was not found. (Miss)

Misnamed Files: Misnamed files (e.g., .txt file renamed with a .dll extension) were not found. (Miss)

Peripheral Memory Cards: N.A. – The Nokia 6610i does not allow for removable media. (NA)

Acquisition Consistency: CellDEK does not provide an overall hash for the case file. However, hashes for individual data elements are consistent after back to back acquisitions. (Meet)

Cleared Devices: N.A. – A Hard Reset function is not provided by the phone. (NA)

Power Loss: The Nokia 6610i was repopulated with the above scenarios, then completely drained of all battery power and reacquired. All data was found as reported above. (Above)

Appendix G: SIM Seizure – External SIM Results

The scenarios were performed on a Forensic Recovery of Evidence Device (FRED) running Windows XP SP2. SIM Seizure version 1.0 build 2468.18222 was used with Paraben's SIM Card Reader to acquire data from a populated SIM.

SIM 534

The following scenarios were conducted using a T-Mobile SIM. Service Provider Name (SPN) was not allocated. No FPLMN entries were registered.

Basic Data: The following data was found and reported: IMSI, ICCID, Language Preference (LP), Abbreviated Dialing Numbers (ADN), Last Numbers Dialed (LND) and active/deleted SMS messages. The ICCID and LP had to be manually decoded for interpretation. ADN entries containing special characters (e.g., '@') were displayed as an empty string. (Below)

Location Data: All LOCI and LOCIGPRS data was found and reported, but had to be manually decoded for interpretation. (Meet)

EMS Data: Active/deleted incoming EMS Messages that exceed 160 characters were found and reported. Messages containing embedded pictures (i.e., 16x16, 32x32 black and white graphics) were found and reported. (Meet)

Foreign Language Data: ADN entries and SMS messages containing French and Asian language characters were found and reported correctly. (Meet)

SIM 877

The following scenarios were conducted using a Cingular SIM. Service Provider Name (SPN) was not allocated. No FPLMN entries were registered.

Basic Data: The following data was found and reported: IMSI, ICCID, Language Preference (LP), Abbreviated Dialing Numbers (ADN), Last Numbers Dialed (LND) and active/deleted SMS messages. The ICCID and LP had to be manually decoded for interpretation. ADN entries containing special characters (e.g., '@') were displayed as an empty string. (Below)

Location Data: All LOCI and LOCIGPRS data was found and reported, but had to be manually decoded for interpretation. (Meet)

EMS Data: Active/deleted incoming EMS Messages that exceed 160 characters were found and reported. Messages containing embedded pictures (i.e., 16x16, 32x32 black and white graphics) were found and reported. (Meet)

Foreign Language Data: ADN entries and SMS messages containing French and Asian language characters were found and reported correctly. (Meet)

SIM 114

The following scenarios were conducted using an AT&T SIM. Service Provider Name (SPN) was not allocated. No FPLMN entries were registered.

Basic Data: The following data was found and reported: IMSI, ICCID, Language Preference (LP), Abbreviated Dialing Numbers (ADN), Last Numbers Dialed (LND) and active/deleted SMS messages. The ICCID and LP had to be manually decoded for interpretation. ADN entries containing special characters (e.g., '@') were displayed as an empty string. (Below)

Location Data: All LOCI and LOCIGPRS data was found and reported, but had to be manually decoded for interpretation. (Meet)

EMS Data: Active/deleted incoming EMS Messages that exceed 160 characters were found and reported. Messages containing embedded pictures (i.e., 16x16, 32x32 black and white graphics) were found and reported. (Meet)

Foreign Language Data: ADN entries and SMS messages containing French and Asian language characters were found and reported correctly. (Meet)

Appendix H: GSM .XRY – External SIM Results

The scenarios were performed on a Forensic Recovery of Evidence Device (FRED) running Windows XP SP2. GSM .XRY version 3.0 was used with the Micro Systemation SIM Card Reader to acquire data from a populated SIM.

SIM 534

The following scenarios were conducted using a T-Mobile SIM. Service Provider Name (SPN) was not allocated. No FPLMN entries were registered.

Basic Data: The following data was found and reported: IMSI, ICCID, Language Preference (LP), Abbreviated Dialing Numbers (ADN), Last Numbers Dialed (LND) and active/deleted incoming SMS messages. (Meet)

Location Data: All LOCI data was found and reported, but had to be manually decoded for interpretation. GPRSLOCI data was not found. (Below)

EMS Data: Active/deleted incoming EMS Messages that exceed 160 characters were found and reported. Messages containing embedded pictures (i.e., 16x16, 32x32 black and white graphics) were found and reported. (Meet)

Foreign Language Data: ADN entries and SMS messages containing French and Asian language characters were found and reported correctly. (Meet)

SIM 877

The following scenarios were conducted using a Cingular SIM. Service Provider Name (SPN) was allocated but not activated. No FPLMN entries were registered.

Basic Data: The following data was found and reported: IMSI, ICCID, Language Preference (LP), Abbreviated Dialing Numbers (ADN), Last Numbers Dialed (LND) and active/deleted incoming SMS messages. (Meet)

Location Data: All LOCI data was found and reported, but had to be manually decoded for interpretation. GPRSLOCI data was not found. (Below)

EMS Data: Active/deleted incoming EMS Messages that exceed 160 characters were found and reported. Messages containing embedded pictures (i.e., 16x16, 32x32 black and white graphics) were found and reported. (Meet)

Foreign Language Data: ADN entries and SMS messages containing French and Asian language characters were found and reported correctly. (Meet)

SIM 114

The following scenarios were conducted using an AT&T SIM. Service Provider Name (SPN) was not allocated. No FPLMN entries were registered.

Basic Data: The following data was found and reported: IMSI, ICCID, Language Preference (LP), Abbreviated Dialing Numbers (ADN), Last Numbers Dialed (LND) and active/deleted incoming SMS messages. (Meet)

Location Data: All LOCI data was found and reported, but had to be manually decoded for interpretation. GPRSLOCI data was not found. (Below)

EMS Data: Active/deleted incoming EMS Messages that exceed 160 characters were found and reported. Messages containing embedded pictures (i.e., 16x16, 32x32 black and white graphics) were found and reported. (Meet)

Foreign Language Data: ADN entries and SMS messages containing French and Asian language characters were found and reported correctly. (Meet)

Appendix I: SecureView – External SIM Results

The scenarios were performed on a Forensic Recovery of Evidence Device (FRED) running Windows XP SP2. SecureView version 1.5.0 was used with a PC/SC SIM Card Reader to acquire data from a populated SIM.

SIM 534

The following scenarios were conducted using a T-Mobile SIM. Service Provider Name (SPN) was not allocated. No FPLMN entries were registered.

Basic Data: The following data was found and reported: Abbreviated Dialing Numbers (ADN). The following data was not found: IMSI, ICCID, Last Numbers Dialed (LND), active incoming SMS message, Language Preference (LP) and deleted SMS messages. (Below)

Location Data: No data was found. (Miss)

EMS Data: No data was found. (Miss)

Foreign Language Data: ADN entries containing French and Asian language characters were found but not displayed correctly. (Below)

SIM 877

The following scenarios were conducted using a Cingular SIM. Service Provider Name (SPN) was allocated but not activated. No FPLMN entries were registered.

Basic Data: The following data was found and reported: Abbreviated Dialing Numbers (ADN). The following data was not found: IMSI, ICCID, Last Numbers Dialed (LND), active incoming SMS message, Language Preference (LP) and deleted SMS messages. (Below)

Location Data: No data was found. (Miss)

EMS Data: No data was found. (Miss)

Foreign Language Data: ADN entries containing French and Asian language characters were found but not displayed correctly. (Below)

SIM 114

The following scenarios were conducted using an AT&T SIM. Service Provider Name (SPN) was not allocated. No FPLMN entries were registered.

Basic Data: The following data was found and reported: Abbreviated Dialing Numbers (ADN). The following data was not found: IMSI, ICCID, Last Numbers Dialed (LND), active incoming SMS message, Language Preference (LP) and deleted SMS messages. (Below)

Location Data: No data was found. (Miss)

EMS Data: No data was found. (Miss)

Foreign Language Data: ADN entries containing French and Asian language characters were found but not displayed correctly. (Below)

Appendix J: PhoneBase2 – External SIM Results

The scenarios were performed on a Forensic Recovery of Evidence Device (FRED) running Windows XP SP2. PhoneBase2 version 1.2.0.15 was used with a PC/SC SIM Card Reader to acquire data from a populated SIM.

SIM 534

The following scenarios were conducted using a T-Mobile SIM. Service Provider Name (SPN) was not allocated. No FPLMN entries were registered.

Basic Data: The following data was found and reported: IMSI, ICCID, Abbreviated Dialing Numbers (ADN), Last Numbers Dialed (LND) and active and deleted SMS messages. The following data was not found: Language Preference (LP). However, the MNC portion of the IMSI, a three-digit value, was incorrectly translated and ADN data elements were not decoded excluding the last entry. (Below)

Location Data: All LOCI data was found and reported. However, the MNC portion of the LAI was incorrectly translated. GPRSLOCI data was not found. (Below)

EMS Data: Active and deleted incoming EMS Messages that exceed 160 characters were found and reported. An EMS message containing an embedded picture was found. However, the image was not correctly decoded and presented, though the text was. (Below)

Foreign Language Data: ADN entries and SMS messages containing French language characters were found and reported. However, ADN entries and SMS messages containing Asian language characters were not displayed correctly in either the user interface or the generated report. (Below)

SIM 877

The following scenarios were conducted using a Cingular SIM. Service Provider Name (SPN) was allocated but not activated. No FPLMN entries were registered.

Basic Data: The following data was found and reported: IMSI, ICCID, Abbreviated Dialing Numbers (ADN), Last Numbers Dialed (LND) and active and deleted SMS messages. No Language Preference (LP) data was found. However, the MNC portion of the IMSI, a three-digit value, was incorrectly translated and ADN data elements were not decoded with the exception of the last entry. (Below)

Location Data: All LOCI data was found and reported. However, the MNC portion of the LAI was incorrectly decoded. GPRSLOCI data was not found. (Below)

EMS Data: Active and deleted incoming EMS Messages that exceed 160 characters were found and reported. An EMS message containing an embedded picture was found. However, the image was not correctly decoded and presented, though the text was. (Below)

Foreign Language Data: ADN entries and SMS messages containing French language characters were found and reported. However, ADN entries and SMS messages containing Asian language characters were not displayed correctly in either the user interface or the generated report. (Below)

SIM 114

The following scenarios were conducted using an AT&T SIM. Service Provider Name (SPN) was not allocated. No FPLMN entries were registered.

Basic Data: The following data was found and reported: IMSI, ICCID, Abbreviated Dialing Numbers (ADN), Last Numbers Dialed (LND) and active and deleted SMS messages. No Language Preference (LP) data was found. However, the MNC portion of the IMSI, a three-digit value, was incorrectly decoded and reported and ADN data elements were not decoded with the exception of the last entry. (Below)

Location Data: All LOCI data was found and reported. However, the MNC portion of the LAI was incorrectly decoded. GPRSLOCI data was not found. (Below)

EMS Data: Active and deleted incoming EMS Messages that exceed 160 characters were found and reported. An EMS message containing an embedded picture was found. However, the image was not correctly decoded and presented, though the text was. (Below)

Foreign Language Data: ADN entries and SMS messages containing French language characters were found and reported. However, ADN entries and SMS messages containing Asian language characters were not displayed correctly in neither the user interface or in the generated report. (Below)

Appendix K: CellDEK – External SIM Results

The scenarios were performed on the CellDEK terminal running Windows XP SP2. CellDEK version 1.3.1.0 was used with CellDEK's internal PC/SC SIM Card Reader to acquire data from a populated SIM.

SIM 534

The following scenarios were conducted using a T-Mobile SIM. Service Provider Name (SPN) was not allocated. No FPLMN entries were registered.

Basic Data: The following data was found and reported: IMSI, ICCID, Abbreviated Dialing Numbers (ADN), Last Numbers Dialed (LND) and active and deleted incoming SMS messages. The following data was not found: Language Preference (LP). (Below)

Location Data: No data was found. (Miss)

EMS Data: Active incoming EMS messages that exceed 160 characters were found and reported. An EMS message containing an embedded picture was found. However, the image was not correctly decoded and presented, though the text was. Deleted EMS messages were not found. (Below)

Foreign Language Data: ADN entries and SMS messages containing French language characters were found and reported. However, ADN entries and SMS messages containing Asian language characters were not displayed correctly in the user interface, although the SMS messages appeared correctly in the generated .rtf report. (Below)

SIM 877

The following scenarios were conducted using a Cingular SIM. Service Provider Name (SPN) was allocated but not activated. No FPLMN entries were registered.

Basic Data: The following data was found and reported: IMSI, ICCID, Abbreviated Dialing Numbers (ADN), Last Numbers Dialed (LND) and active incoming SMS messages. The following data was not found: Language Preference (LP) and deleted SMS messages. The initial portion of the IMSI was incorrectly appended to the reported ICCID. (Below)

Location Data: No data was found. (Miss)

EMS Data: Active incoming EMS messages that exceeded 160 characters were found and reported. An EMS message containing an embedded picture was found. However, the image was not correctly decoded and presented, though the text was. Deleted EMS messages were not found. (Below)

Foreign Language Data: ADN entries and SMS messages containing French language characters were found and reported. However, ADN entries and SMS messages containing Asian language characters were not displayed correctly in the user interface, although the SMS messages appeared correctly in the generated .rtf report. (Below)

SIM 114

The following scenarios were conducted using an AT&T SIM. Service Provider Name (SPN) was not allocated. No FPLMN entries were registered.

Basic Data: The following data was found and reported: IMSI, ICCID, Abbreviated Dialing Numbers (ADN), Last Numbers Dialed (LND) and active SMS messages. The following data was not found: Language Preference (LP) and deleted SMS messages. The initial portion of the IMSI was incorrectly appended to the reported ICCID. (Below)

Location Data: No data was found. (Miss)

EMS Data: Active incoming EMS messages that exceed 160 characters were found and reported. An EMS message containing an embedded picture was found. However, the image was not correctly decoded and presented, though the text was. Deleted EMS messages were not found. (Below)

Foreign Language Data: ADN entries and SMS messages containing French language characters were found and reported. However, ADN entries and SMS messages containing Asian language characters were not displayed correctly in the user interface, although the SMS messages appeared correctly in the generated .rtf report. (Below)

Appendix L: USIMdetective – External SIM Results

The scenarios were performed on a Forensic Recovery of Evidence Device (FRED) running Windows XP SP2. USIM Seizure version 1.3.1 was used with a PC/SC SIM Card Reader to acquire data from a populated SIM.

SIM 534

The following scenarios were conducted using a T-Mobile SIM. Service Provider Name (SPN) was not allocated. No FPLMN entries were registered.

Basic Data: The following data was found and reported: IMSI, ICCID, Language Preference (LP), Abbreviated Dialing Numbers (ADN), Last Numbers Dialed (LND) and active/deleted SMS messages. (Meet)

Location Data: All LOCI and LOCIGPRS data was found and reported. (Meet)

EMS Data: Active/deleted incoming EMS Messages that exceed 160 characters were found and reported. Messages containing embedded pictures (i.e., 16x16, 32x32 black and white graphics) were found, but the images were not displayed. (Below)

Foreign Language Data: ADN entries and SMS messages containing French characters were found and reported. Messages and ADN entries containing Asian characters were not presented correctly. (Below)

SIM 877

The following scenarios were conducted using a Cingular SIM. Service Provider Name (SPN) was not allocated. No FPLMN entries were registered.

Basic Data: The following data was found and reported: IMSI, ICCID, Language Preference (LP), Abbreviated Dialing Numbers (ADN), Last Numbers Dialed (LND) and active/deleted SMS messages. (Meet)

Location Data: All LOCI and LOCIGPRS data was found and reported. (Meet)

EMS Data: Active/deleted incoming EMS Messages that exceed 160 characters were found and reported. Messages containing embedded pictures (i.e., 16x16, 32x32 black and white graphics) were found, but the images were not displayed. (Below)

Foreign Language Data: ADN entries and SMS messages containing French characters were found and reported. Messages and ADN entries containing Asian characters were not presented correctly. (Below)

SIM 114

The following scenarios were conducted using an AT&T SIM. Service Provider Name (SPN) was not allocated. No FPLMN entries were registered.

Basic Data: The following data was found and reported: IMSI, ICCID, Language Preference (LP), Abbreviated Dialing Numbers (ADN), Last Numbers Dialed (LND) and active/deleted SMS messages. (Meet)

Location Data: All LOCI and LOCIGPRS data was found and reported. (Meet)

EMS Data: Active/deleted incoming EMS Messages that exceed 160 characters were found and reported. Messages containing embedded pictures (i.e., 16x16, 32x32 black and white graphics) were found, but the images were not displayed. (Below)

Foreign Language Data: ADN entries and SMS messages containing French characters were found and reported. Messages and ADN entries containing Asian characters were not presented correctly. (Below)

www.ingramcontent.com/pod-product-compliance
Lightning Source LLC
Chambersburg PA
CBHW081723170526
45167CB00009B/3683